RACING TO EXTINCTION

RACING TO EXTINCTION
WHY HUMANITY WILL SOON VANISH

Lyle Lewis

ENDANGERED PRESS

SPRINGFIELD, OR

Racing to Extinction
Endangered Press, Springfield, OR

© 2023 by Lyle Lewis

Book design by Vinnie Kinsella, Paper Chain Book Publishing Services.

ISBN (paperback): 979-8-9896381-0-9
ISBN (ebook): 979-8-9896381-1-6
ISBN (audiobook): 979-8-9896381-2-3

Library of Congress Control Number: 2023923308

DISCLAIMER

This is a work of nonfiction. The current health and status of the planet has not been embellished. No names have been changed. The current and past actions and behavior of *Homo sapiens* have not been fabricated. No characters have been invented. The author is not responsible in any manner whatsoever for any mental challenges that may arise from readers absorbing the information provided in this book, including but not limited to anger, depression, horror, alarm, and panic.

To my friends who died too young. Their sharp and inquisitive minds, kind hearts, unflagging support, and sense of humor made my life happier and more fulfilling.

Patti Pansa
Charles "Chuck" Harris
Jo Ellen Wenger
Allen Thomas
Elizabeth "Dixie" Pierson
David Campbell

CONTENTS

FOREWORD

The writing has been on the wall for generations, and in each generation, voices were lifted in an attempt to direct the attention of the general public and those in power to the consequences of continuing to ignore the writing on the wall.

The Romantic poets saw and felt the horrors of early-stage Industrialisation and its dysfunctional social order because they had hearts and their eyes were not blinded by the contemplation of money and profit. They lifted up their voices in lament, as did some of their visual artist colleagues, like J. M. W. Turner, and poets coming after, including Walt Whitman and e. e. cummings.

Charles Dickens lifted up his voice at the cut-throat cruelty of his own society, which was at that point also busy colonising the rest of the world, exporting its cruelty abroad. His Magwitch speaks for all abused children, trampled citizens, and hunted men in history.

Indigenous people like Chiksika lifted up their voices in protest against the destruction that their foreign oppressors were wreaking on their and our Mother. They still lift up their voices today when industry comes to what is left of their traditional lands to, e.g., frack for gas, lay oil pipelines, annihilate wildlife and habitat for lithium mines, and bulldoze sacred sites like Australia's Juukan Gorge or Nevada's Thacker Pass. Few in mainstream society hear them, and even fewer care.

Concerned scientists and other citizens have been lifting up their voices in warning for centuries about the way we are treating this planet. They have spoken out about habitat destruction, human overshoot, pollution—and are mostly met with indifference or disdain by mainstream society, especially the upper echelons that have grown rich and powerful from their exploitation and abuse of nature and their fellow humans.

The Earth's biosphere is disintegrating, with the vast majority of humans either in great ignorance or in denial. The dominant, toxic culture that sees our brother and sister species—and our brother and sister humans—as resources to be consumed and capital to be exploited has grown into a physically voracious, planetwide machine that is destroying ecosystems, meaningful human communities, and human and nonhuman lives day and night and at ever-accelerating rates.

The machine was once smaller, powered by oxen, horses, donkeys, human slaves, the burning of the bodies of trees. The first empires rode on its back, including the famous Roman Empire, to conquer and steal, to rape and pillage, amongst other human and nonhuman communities, in order to enrich themselves at others' expense. The faces kept on changing, but the plot remained the same. With the Industrial Revolution, the machine became driven by engines and wage slaves, and soon most of the engines ran on fossil fuels, and with the energy thus unleashed, the destruction skyrocketed.

Many people now understand that the burning of the planet's fossil carbons causes carbon dioxide to accumulate in the atmosphere, with consequent planetary warming and climate change. Few see that the opposite side of this equation—the ability of plants on this Earth to sequester carbon—has also been severely disrupted by the continued human-orchestrated destruction of global ecosystems. The combined effect is that the climate is now leaving the stable Holocene period and entering uncharted territory.

The destruction of the planet's web of life is a screaming-ambulance-siren problem not merely because it is the other main driver of our climate crisis but in and of itself. Nobody should have to point out that it is appalling to lose a *single species* to human activity, let alone the current reality of over 150 species *a day*. Nobody should be numb to such a biological apocalypse, to such wholesale killing caused not by somebody else *over there* but by each and every one of us living in the toxic, genocidal, ecocidal machine we call *civilisation*.

Civilisation has always in practice meant urbanisation. People living in cities have to import their food and other necessities (and luxuries) from outside the city, traditionally from its hinterland, and since globalisation, from all over the planet, at even greater ecological expense. When the population of the city grows, when people want more stuff, or when the hinterland necessarily degrades—and usually, all three apply simultaneously—the city needs to either increase the intensity of extraction of its existing hinterland, which will further degrade it ecologically and often exacerbate social injustice, or acquire more hinterland. The latter means dispossessing other people and/or wiping out more wildlife habitat—in short, genocide and ecocide.

Ecosystem destruction looks like this: deforestation, logging, clearing for the usual reasons of agriculture, suburban expansion, mining, road building and industrial projects including power transmission lines and "green" solar and wind farms; also human-induced land degradation/desertification and the ongoing toxification of the planet by industrial and household chemicals and plastic pollution. These are human crimes against life on Earth, and they are baked into the system in which most of us are currently organised, each of us little cogs in the destructive machine of contemporary consumer society.

We who are heirs to this destructive culture should be repenting in sackcloth and ashes at the carnage and desecration we are causing on this planet and doing our level best to dismantle our destructive machine and reintegrate with the natural world we have spat upon and trodden under our feet for so long, instead of pretending that tinkering around the edges of the machine is a sufficient response. Our level best includes voluntarily limiting family size to counter our vast population overshoot, wiping out social inequality and the hierarchies behind it, *drastically* reducing average personal consumption, dismantling our large-scale industrial infrastructure, ending the profligate waste of consumerism and globalisation, phasing out human feedlots, returning to simpler smaller-scale technologies, making healthy supportive communities, engaging hands-on with ecologically sound revegetation efforts and with growing our own food in permaculture systems, and striving for local self-sufficiency.

Cue predictable screaming from the military-industrial complex, the corporate sector, and the politicians and media owned by them. Also from local chambers of commerce and entitled people anywhere—including in the field of science—for whom the status quo, privilege, and personal freedom are not only more important than other people's health and having fair and functional human community but infinitely more important than reversing the frenzied assault of our species on all remaining wild forms of life—on the very web of life that supports all life on this planet, and from which we are removing more strands 24/7. Cue additional screaming from those in the general population—*and* in the field of science—who do not understand what we are doing to the biosphere collectively, why this is an atrocity, and how it will come back to bite everyone.

Drastic change should have happened 50, 100, 200 years ago, when the writing was already on the wall and some thinkers and poets in every generation could see it, just as many Indigenous people could see it. This has now become a palliative care exercise. The house of cards *Homo colossus* has been busy building for centuries is collapsing, and the sixth great extinction that is underway courtesy of our ecocidal activites is poised to sweep us off the stage with it.

It's funny how we seem to feel immortal when we are young, as if disease and disaster can't possibly happen to us. Likewise, it's funny how as a culture we seem to feel immortal, as if no matter how much damage we do to the natural systems that support life on Earth, and no matter how many other species we send to extinction daily, it can't possibly happen to us. It's also funny how most people regard the *idea* of human extinction with an abject horror they are unable to summon for our brother and sister species.

One of the biggest wake-up calls for me personally was finding out as a young scientist and educator that it's not just the corporate sector, and the politicians who dance to their handclaps in the name of "jobs and *growth*," aka ecocide, who are making it systematically almost impossible for concerned citizens to effect large-scale necessary change. It is also that the very agencies charged with the protection of the environment, with the education, health, and welfare of humans, and with justice in society, are complicit in undermining all of these instead.

Lyle Lewis also found this out. He too was a caring young person who genuinely wanted to make a positive difference to wildlife and the health of the biosphere by studying biology and then working for government agencies charged with nature conservation. He spent 30 years working for such agencies, and when I read about the experiences he relates in this book, I nodded from the start to the finish of his recount. I'd seen all the things he described myself, in Australia. Many colleagues I have talked to, who started out as caring young people excited about going in to bat for nature, have also seen these things and have experienced the disgust and despair of witnessing such systemic incompetence, obfuscation, and corruption.

Our stories do not get mainstream press. Most people do not want to hear this. Our voices are routinely silenced. In this book, Lyle speaks for all of us who studied and loved the Earth and wanted to make a difference through our chosen field of study and the bureaucracies allocated to nature conservation by our vastly dysfunctional, nature-divorced society—and found out along the way how that goes down.

Another person who found out how environmental truth-telling goes down with the establishment was Rachel Carson, when she began to question the safety of DDT back in the 1940s. She hit brick walls trying to make her concerns public, but persisted and in 1962 published *Silent Spring*, with the result that she was hounded by the chemical industry and its shills in government and other positions of power. She was ridiculed for being a "bird and bunny lover" and not having a PhD or an academic position, and with typical establishment misogyny, described as "hysterical" and a "spinster" and unstable and acting beyond her qualifications. Rachel Carson was right about DDT, though, and unlike her pathetic accusers, had decency, integrity, and joined-up thinking.

Truth-telling isn't just unpopular with the establishment, but also, too often these days, with the public. Nine years before Pope John Paul II publicly admitted and apologised for the Catholic Church's systemic paedophilia cover-ups, Sinéad O'Connor, a survivor of childhood sexual abuse, in 1992 famously tore up a picture of the Pope on US television to draw attention to the widespread sexual abuse of children by the Catholic Church. That very picture hung for years in the home of her abuser, yet it was she who became the public's favourite pariah for making a stand on this issue. People did not want to hear. But she was right. Children were abused, and very few perpetrators, even to this day, were ever held accountable for the abuse and its cover-up—including Australian Archbishop George Pell, who died in the Vatican, safely away from Australian victims.

These are not isolated instances of evil or "misunderstanding"—this is standard operating procedure for our toxic hierarchical systems. This is *why* few paedophile priests or rapists in high-status positions are brought to justice and their victims are shredded and slandered instead; *why* decades after alleged gender equality, women are *still* dealing with inequality, discrimination, disrespect, and sexual violence; *why* Indigenous people experience inequality, blatant racism, poorer health, lower life expectancies, etc., and their dispossession never ended; *why* we still have homeless people and working poor while billionaires live it up; *why* politicians and CEOs have vastly higher salaries than nurses or teachers, even though their roles are no more demanding or important; *why* a million other injustices exist and persist. No matter how many Parliamentary Inquiries or Royal Commissions are staged and special reports and recommendations are produced, *this never fundamentally changes*.

They say that when someone shows you who they are, believe them. Likewise, I say, when a system shows you what it is, believe it. The system under which we live in our ostensibly modern, wealthy, democratic societies is rotten to the core, and if you open your eyes and look around without your mainstream-issue rose-tinted spectacles, you see it everywhere. We live in a dire dystopia underneath a veneer cleverly constructed by Screwtape and his marketing department, with lots of soma and unicorn dust splashed around. We are brainwashed and indoctrinated from early childhood, like children born into any cult; it is difficult to break away and see it, and painful, and guaranteed to get you ostracised. We have our Two Minutes Hate orchestrated for us and horizontal hostility avidly fanned by the sociopaths who are meanwhile dominating ordinary citizens from atop their greasy poles, and skimming all the cream off the top for themselves and their ilk.

If you critique a rich democracy that drip-feeds consumerism to its populace, you will be invited to look at the poverty, chaos, and corruption in developing countries and to be grateful for your lot, conveniently without deconstructing why the poverty, chaos, and corruption exist in those countries in the first place. It's a similar move as some Western parents invoking starving Ethiopians if their children don't want to finish their plate. It's cynical manipulation and doesn't address the problems on either side, let alone actually care.

Rachel Carson lived to see her concerns about the consequences of DDT use on human health and wildlife backed up by subsequent studies and its use in her own country begin to be legislated so that human communities at least could start to defend themselves from aerial spraying. Shortly after her death (in her 50s, from cancer), DDT was largely banned in the US. But it is still being used even today to kill mosquitoes for malaria control, and as a result of past and ongoing use and its long half-life, the entire globe is contaminated with it. DDT is present in the bodies of all humans even before birth, and in the bodies of wildlife everywhere, even on remote islands where it was never directly sprayed.

Of course, DDT is only one of many dangerous and persistent industrial chemicals found in every organism tested anywhere on the planet, and most of them are far less regulated. PFAS, for example, are another "forever chemical" found in common items including raincoats, Teflon cookware, household stain protection sprays, and food packaging and are routinely dumped on burning buildings and wildlands in the form of firefighting foam. PFAS are now ubiquitously found in household dust and rainwater, and contaminating the bodies of humans and wildlife. Fire retardants are another Pandora's box; they come without warning in your sofas, mattresses, impregnated into children's pyjamas, etc. Plastics and plasticisers are not the inert, harmless substances they were initially claimed to be, and are now filling up the oceans.

In summary, there are hundreds of thousands of industrial chemicals in use, many of which aren't harmless, and few of which are usefully regulated. Thousands of new ones are introduced each year. They come to your home in everyday consumer products—furniture, clothing, carpet, cookware, toiletries, tea bags, packaging. They contaminate the planet at the point of manufacture, eventual disposal, and everywhere in-between, and if they have adverse health or ecological effects, the onus is largely on the public to agitate to get them regulated and to make sure their replacements aren't equally problematic. Meanwhile, every passing day, the planet and its biota become more contaminated with industrial chemicals than ever before.

And Rachel Carson, who loved this Earth, believed people had a right not to be poisoned in their own homes, and dedicated so much of her life to educating the public about nature, ecology, and environmental and public health issues like this, is turning in her grave.

* * *

With this book, Lyle Lewis is lifting up his voice on behalf of the Earth he deeply loves, his brother and sister species whom he deeply loves, and all other people who deeply love the Earth and its magnificent, miraculous, beleaguered life forms. He has my love and my gratitude for caring, for wanting to know, for speaking out, for having a heart, for his integrity and his backbone. I thank him for writing this book. I know he wishes his conclusions could have been different.

In the pages that follow, Lyle takes you on his journey grappling with dark times. He takes a square look at the current state of the planet and at the main drivers of the accelerating ecological disintegration on Earth. He discusses human overshoot, the toxification of the biosphere, climate change, biodiversity loss, and related matters, along with aspects of the dominant human culture and human psychology that are creating and enabling our devastation of Planet Earth. The author has a special interest in human evolution and in the ecological consequences of the spread of our hunter-gatherer ancestors out of Africa and all around the world. He believes that we started making trouble for the wider biotic community well before the Agricultural or Industrial Revolutions, and he shows us how.

Lyle also discusses at length the mechanisms and extent of the postcolonial ecological devastation of North America, in which the agencies he worked for were and remain complicit. He then shares insights into processing the trauma and grief of living in a collapsing biosphere—something those of us who have been witnessing it on the ground are all finding difficult.

As eco-realists, rather than techno-optimists or religious believers in upcoming divine intervention, many of us are consciously working through grief over the daily devastation of our brother and sister species and their steady disappearance from this planet. Many of us are also doing our level best to help them hang on in our local areas, by personally protecting and stewarding remaining patches of native ecosystem and planting native species back into devastated areas to make corridors and refuges useful to wildlife.

The song of a single bird is more eloquent and important than all the chatter of the many "important" people staging useless summits, leading countries, selling snake oil, etc. It is a balm to ears that have heard so much emptiness.

So if you ask us why we still bother to do so given our acceptance that we are living through an accelerating mass extinction, one main reason is that the here and now is all we ever had and is still meaningful, for all forms of life. The other is that we are fighting to leave as much biological heritage as we can for nature to do whatever it will do when we fall on our own sword and stop being the worst plague that ever existed—which is sadly what we are, collectively, in spite of the good and beautiful things that humans are also capable of.

In our Titanic's extensive well-stocked library is a small shelf containing volumes that elucidate with honesty and joined-up thinking the holocaust that modern humans have created for the community of living things on Planet Earth, and which is coming back to bite us.

Lyle Lewis's *Racing to Extinction* sits squarely and deservedly on that shelf.

—SUE COULSTOCK, biologist/environmental scientist, educator, writer, off-grid conservationist and permaculturist

INTRODUCTION

The human population will continue to increase until it can't. When it can no longer increase, it will crash. Our extreme efforts to focus our minds elsewhere are symptoms of a desperate attempt to find solid footing, to believe in a future that will not vanish.

Mankind has had a storied existence on Earth. Our thirst for knowledge and innovation has provided comfort and security. However, it is becoming increasingly evident that our remarkable progress has become our worst enemy.

We know our planet is finite. Images of Earth from space make it plain for all to see. However, people routinely ignore this verifiable certainty in all aspects of their lives, treating the world as infinite. Modern-day human activities are not only wiping out ecosystems and biodiversity but plundering the clean air, water, and topsoil that helped bring about our tenure on the planet. Entire ecosystems have vanished, including the tallgrass prairie in North America, Madagascar's rainforests, and the Aral Sea in Asia. We use Earth's natural resources like a bunch of drunks on the greatest bender of all time. Human consumption is negatively modifying the planet and permanently damaging the biological systems upon which our continued existence depends. The deterioration of our ecosphere has been exponentially accelerating for at least 100,000 years.

As technological advances improve our lives, humanity becomes increasingly detached from its environment and the natural resources allowing us to persist. People now have a much closer affinity with iPhones, Amazon, online shopping, restaurants and bars, Netflix, beauty salons, and sporting events than forests, grasslands, marshes, and oceans.

Few now understand our existence on Earth is entirely dependent upon photosynthesis. Instead, they believe their survival is contingent on parents, doctors, farmers, governments, bankers, police, and other players in society. It's not that those institutions, people, and specialties aren't important, but they represent the retailers. Photosynthesis is the wholesaler. There is a supply chain disruption occurring on a massive scale in our relationship with the planet. The ancient forests and grasslands that provided the planet with free oxygen in the air and sequestered carbon dioxide, making it habitable for humans and other complex life, are nearly gone.

For most people, the natural resources that support their existence and lifestyle might as well be from a distant galaxy. This extreme disconnect has resulted in people losing their capacity to understand the dire circumstances facing complex organisms on Earth, including themselves. The deafening alarm bells portending our extinction are routinely misunderstood or ignored.

For those who perceive our state of crisis, there is a great deal of angst. Much of the frustration, anger, and sadness results from having unrealistic expectations of human beings. Despite our advanced technologies, the basic tenets of human behavior haven't changed for centuries or millennia. Letting go of the false expectations that *Homo sapiens* can or will modify our behavior can bring us an element of peace. Expectations are incredibly powerful in structuring our moods and emotions. Identifying unrealistic expectations can reduce our chances of being disappointed and can increase happiness. A better understanding of our behavioral history can provide valuable insights into recognizing human capabilities and limitations.

When I was growing up on a small ranch in Southern Oregon in relative isolation, nature became my best friend. During my professional career, I held diverse positions including range conservationist, watershed specialist, hydrologist, fisheries biologist, ecologist, wildlife biologist, endangered species branch chief, and endangered species recovery coordinator. More than two decades of my federal career were spent in efforts to conserve imperiled species in the face of livestock grazing, logging, oil and gas development, agricultural development, mining, motorized recreational use, highway expansion, construction, rights-of-way, and urban development.

My varied background, coupled with my rural upbringing, made it possible to examine both environmental and social issues through the lens of multiple disciplines and viewpoints. This unique life experience provided me with an appreciation for the degree to which everything on our planet is interrelated and interconnected. Some described me as thinking outside the box, but more aptly, my perspective was less confined by myths, social constructs, or a single discipline.

My professional life played out on a thin high wire: advocating for wildlife and wild places while trying to mollify supervisors and others using these things for economic growth. Bullying, conflict, and criticism followed me around like playful puppies my entire career. Anyone making sincere attempts to conserve natural ecosystems and biodiversity was viewed as an irritating impediment to growth and prosperity. There was intense pressure to conform to the natural resource management agency's underlying goal of avoiding conflict at all costs.

Despite the never-ending conflicts and occasional pitfalls, I was both lucky and relatively adept at navigating the bureaucratic minefield, spending the last 11 years of my career in low-level management positions. What I acquired during decades of experience, and hope to convey in the subsequent chapters, is a heightened understanding of the interface of landscape ecology, evolutionary biology, and human psychology.

Because I'm not a specialist, there are many who know more about wildlife, forests, grasslands, oceans, and wetlands. I'm not an epidemiologist, paleontologist, climatologist, toxicologist, psychologist, economist, geologist, statistician, or astronomer. I'm not an expert in history, theology, physiology, human anatomy, mathematics, physics, politics, genetics, or sociology.

However, all these disciplines are important variables in an equation explaining the current and future status of *Homo sapiens* and our planet. Specialization is now a firmly entrenched part of modern society. Specialists deserve our respect and admiration, and they have mine. However, our complete reliance on specialists and aversion to integrating seemingly disparate fields will ensure humanity wanders myopically into our limited future, careening from one crisis to the next. This book seeks to weave each of these important fields of study into a tapestry that delivers a colorful and unified portrait of the history, current status, and future of humanity, other organisms, and the planet.

In my relatively small circle of friends and acquaintances, nearly a quarter of them have independently told me in private conversations that they don't think humanity can survive. They are representative of a small but increasing segment of society who understand that our impact on the planet isn't commensurate with its capacity to support humanity. This book gives voice to those increasingly loud whispers and private conversations.

There is a fine line between pessimism and realism. The line nearly disappears when looking at humanity's relationship with the planet. Dreadful things are happening. A realist is someone who makes completely unbiased judgments and doesn't perceive their world through any kind of filter. As a result, true realism is a mirage because everyone filters information.

Every situation bears a million different details to consider and just as many memories to compare. There is not only too much information for an individual to process but also too much for humanity. This book examines the world through the lens of my experiences and observations. Simultaneously, I attempt to step even further out on a fragile limb by assimilating information from a great many disciplines, each with a nearly infinite amount of information. After reading this book, readers will be better positioned to determine through their own filters the status and future of humanity and the planet.

The upper management of state and federal natural resource agencies in most governments are largely populated by individuals whose primary concern is political and economic expediency at the expense of wildlife, native plants, and their habitats. These individuals largely do what politicians are asking of them. Similarly, politicians are doing what huge swaths of the public and wealthy corporate sponsors ask. The only ones that don't benefit from this arrangement are wildlife, wildlands, their advocates, and the planet. Native plants and animals have and will continue to suffer incredible losses until humanity disappears and the planet reboots the evolutionary process.

Living contemporaneously and happily in the final days and years of humanity's existence on the planet is a serious mental hurdle. Happy, positive people who fully understand the consequences of what humanity is doing to the planet are rare. This book will help readers discern between those people who are uplifting, successful, and using global collapse for economic gain from those who are keenly aware of our negative incursions on the planet. I also hope to help you find increased happiness, by minimizing unrealistic expectations of the capabilities of humanity.

Psychologists and motivational speakers advise surrounding yourself with genuine, affirmative people and avoiding people with a negative message. This advice is a catch-22. There is an increasingly obvious, unpleasant reality enveloping Earth. Many who either ignore or discount our effects on the planet are ideologues of either one or multiple fantasies. This allows them not only to disengage from the ecological reality but to encourage others to do the same.

We all share culpability in what is transpiring. Even though there are people contributing much more than others to the problems that will precipitate our demise, no one is beyond reproach. The adage is apt: those who live in glass houses shouldn't throw stones.

Many of those actively seeking to destroy the planet are nice people in every other way. They're the same folks you'd meet in restaurants, bars, sporting events, churches, fairs, and gyms. They represent the entire spectrum of people in modern-day society. It would be easier to fight threats to our limited tenure on the planet if these people were monsters. Some are; most are not.

As with everything biological and behavioral, exceptions abound. This book describes many biological and behavioral concepts having exceptions.

If you are reading this book, you too are likely an exception to the general population, if for no other reason than having an abject curiosity far exceeding your counterparts. If you are human, you're probably guilty of pondering some of the same questions this book addresses.

My experiences working with knowledgeable individuals regularly studying and conserving fish, wildlife, and plants have revealed that those who often know the most realize their knowledge is incredibly superficial. They understand clearly that every query answered spawns dozens more questions. The universe of our existence is much more complex than we have the intellectual capability or technology to understand. Our ignorance of the natural world is a boundless resource. This self-awareness is similar to the humble doctor who admits that the more he learns, the stupider he feels.

With the advent of the internet, the modern world is inundated with information. With the advent of AI, the world will soon be drowning in even more misinformation. Increasingly, the value of experts lies in their capacity to disregard extraneous information and extract meaning from that which is relevant.

This book covers the darkest topic imaginable. Extinction is both horrible and final. Human extinction is ghastly. Even those people who have accepted humanity's plight don't normally spend time scrutinizing the reasons because the subject matter is too painful.

How many of us can bear to hear one sobering bad news story after another? For that reason alone, examining this topic in detail will automatically be viewed by many as callous and insensitive. Making a silk purse out of a sow's ear just isn't possible though.

Confirmation bias is integral to many endeavors. In order to write a book about any topic, an author must seek supporting information for their chosen subject matter. It would normally be a challenge for this inherent bias not to blind the investigator to alternative theories. While writing this book, I sincerely wished to uncover information that would disprove my bias.

Secreting away into the catacombs of a university library for obscure data and text to support my confirmation bias was completely unnecessary. Nearly all the information not only was readily available, but there were times when the book was seemingly writing itself. Barely a day passed when current events or new data didn't provide additional supportive information relevant to this book.

Capitalism, socialism, or any other form of government all depend to varying degrees on the rewards of innovation. Without this, economies, mental thought, and even evolution cannot move forward. We can neither escape it nor flee its detrimental effects.

The human propensity to use whatever natural resources are on hand to increase prosperity, comfort, and wealth and dispassionately drive other organisms to extinction, has now been a recurring theme for well over one hundred thousand years. My experiences—resulting from a lifetime of natural resource management and interactions with a diverse array of people who directly influence the habitability of our planet—aren't unique or new.

There has been an inevitability to the sixth mass extinction now transcending several million years. Other than being the species precipitating the current extinction event, we're nothing special. We're another species doing what all species do. Does that mean we close our eyes and pretend it isn't happening? For the majority of the people on the planet, the answer to that question is a definitive yes. Whether they are aware of it or not, every person on the planet makes decisions each day affecting how rapidly extinctions occur, including our own.

Irrespective of whether a person considers themselves liberal or conservative, this book will challenge conventional wisdom and beliefs on a variety of issues. The information provided would be made most useful by readers setting aside preconceived notions and beliefs and considering my observations and experiences with a critical eye. It is my hope that this book resonates with individuals willing to ask awkward questions, trace difficult connections, and challenge societal dogma. The following chapters are meant to provide the reader a panorama of the convergence of humanity and our planet in both time and space.

ANOTHER DAY

Just as morning light was ushering in a new day 1.4 million years ago, a small group of hominids emerged from a shallow cave in a steep ravine. Without a sound, they surveyed the landscape. A small, slow-moving river downslope from the group was a ribbon of tranquility, winding its way into a lake several kilometers away in what is now Kenya. The water was clear and cool, flowing with a gentle murmur as it meandered along its path. Three giant buffalo, slightly larger than modern-day Cape buffalo and with longer horns, could be seen grazing along the river. A horse-like gemsbok with its distinctive tan, black, and white facial features and sporting long, straight twin horns was nibbling the tops of grasses on a large plateau-savanna across the ravine. Upstream was a wetland created by an oxbow in the river. Occasionally breaking the orange-hued horizon were the tops of toadstool-like acacia trees.

It had rained during the night, and there was a pleasant earthy smell mixed with that of the gardenia bushes and wild basil scattered across the slope of the ravine. Hornbills and a kingfisher could be heard calling from the trees and brush along the river. A fish eagle flew toward the lake.

For five minutes, none of the small band moved. Clothed with short skirts made from animal skins and armed with wooden spears, they surveyed the skies and listened carefully for noises that might indicate danger or a source of food nearby. The sunrise was like a slow bleed of orange and pink across the sky, spreading outward in a canvas of warmth. Finally, satisfied there was no immediate danger and with the sun barely peeking over the horizon, one of the women nodded toward the cave entrance. Two children materialized from the darkness of the cave, one about four years old and the other 11, and joined the adults who ranged in age from 20 to 65.

Using the blunt end of their spears as walking sticks, the seven men, four women, and two children moved down a well-worn trail through brush and lemon grass to the small river where they alternated drinking water and standing watch. Thirst satisfied, the group walked downstream about 30 minutes to the small delta where the stream entered the lake, watching their back trail carefully. The buffalo they had seen earlier had moved onto the savanna, so other than a brief glimpse of an elephant shrew preying on insects, they saw little before the lake came into view. The lake was a mirror, reflecting the deep blue of the sky and brown of the landscape. The calm surface was disturbed by the occasional gentle ripples caused by fish breaking the surface. A small herd of sable antelope were drinking at the shoreline a kilometer away.

The oldest man climbed a steep rock face near the river delta. Perched between two large boulders, he had an unobstructed view of the surrounding area to detect any approaching danger while being easily visible. The group had long ago refined arm signals that would allow them to communicate immediately with each other. After watching the water for disturbances indicative of either danger or food and seeing only the helicopter-like orange dragonflies common to the area, the entire group waded into knee-deep water searching for mollusks and to hunt fish holding or swimming in shallow water. Over the next 45 minutes, one of the women speared a fish, and the group collected several dozen mollusks attached to the surface of underwater rocks. Suddenly, the lookout gave an arm signal for imminent danger. Everyone ran for shore just prior to a pair of large Nile crocodiles suddenly materializing just a few meters from where the group had recently foraged.

The group walked upriver a short distance where the lookout joined them for a clam and fish breakfast. The fish was only large enough to provide a bite or two for each person. The clams were quickly extracted with a rock flake, distributed, and eaten. Unable to resume foraging in the shallow lake environment because of the lounging crocodiles, the humans quenched their thirst again from the river before walking north along the lakeshore. The group knew of a patch of horned melon ripening from previous forays and after a two-kilometer hike began gathering the cucumber-like fruit. After picking and eating the small melons for a half hour, the group hiked up to a small flat near an acacia tree where they could rest; they could climb into the tree canopy should danger appear unexpectedly.

An hour later, one of the children spotted several vultures circling an area near their cave. The group immediately began retracing their steps from the morning until they were a half of a kilometer downstream from the birds. One of the group hiked up a ridge in the ravine where she could view the area below the spiraling avian scavengers. She signaled danger ahead and to hunker down. The scout returned to the group, taking care to remain out of sight of whatever peril she had seen. The group communicated with hand signals, whispers, and drawings in the dirt to remain undetected by a giant otter somewhat resembling a bear that had killed an eland calf near the river oxbow-wetland upstream of their cave. Eland are the largest antelope on Earth, and the four-month-old dead calf weighed nearly 80 kilograms.

Scattered African red alder trees along the river obscured the bear-otter and the hominids from each other's view. A downriver breeze prevented the predator from detecting the approaching humans. The decision was made for the entire group to close the distance, and when a hundred meters from the carnivore, a man and both children would move up a ridge to a vantage point where they could maintain surveillance for additional danger.

As the six men and four women neared the bear-otter, they could hear bones breaking as the large predator fed on the eland calf. The carnivore realized it had company when the group appeared about 40 meters away from behind a copse. About two meters long with dark-brown fur and weighing nearly as much as an African lion, the otter had a head similar to a black bear. The bear-otter was a formidable animal of great beauty and grace, and its fur was dark and sleek. Its eyes were bright, alert, and piercing, seeming to see into the very soul of the Earth. It was a creature of the water and the land, a being that could swim with the ease of a fish and run with the swiftness of a leopard. Bear-otters were the largest member of the weasel family to have ever walked the planet.

The perturbed animal immediately snarled and bluff-charged the group. The humans, moving quickly and carefully, and hearts pounding with fear and adrenaline, formed a wedge with their two-meter-long spears pointing outward. The powerful predator was incredibly quick for its size, but the humans had decades of practice working together in similar situations. The humans yelled and thrust their spears at the enraged animal. For ten minutes, the bear-otter and the humans went back and forth in a standoff battle as the animal tried to protect its kill from the scavengers. At one point, the bear-otter bit the end off one of the spears, but one of the men quickly stabbed a spear into its foreleg. The group quickly reformed their wedge to prevent the slightly wounded animal from exacting revenge. The small clearing of grass and sedges near the river was soon trampled flat by the ongoing skirmish.

Centimeters at a time the humans gained ground on the calf carcass. The bear-otter snarled and bluff-charged time after time, but the group of humans rebuffed each assault finally reaching the partially eaten calf. The woman who had lost her spear quickly grabbed the leg of the calf and began dragging it downstream, no easy task for someone weighing less than the dead animal. The rest of the group held their ground against the furious predator, protecting the woman, and then slowly backed away in her wake. The infuriated carnivore continued attacking, each time rebuffed by the well-practiced thieves. Finally giving up, the slightly wounded animal climbed out of the ravine onto the savanna in its never-ending quest for food and survival, a symbol of the raw power of nature.

The two children rejoined the larger group, but the man stayed at his lookout remaining vigilant for danger. Already two yellow jackals were shadowing the group, hoping for scraps. The number of white-backed vultures circling overhead had increased from three to a dozen just in the short time since the bear-otter had killed the calf. The humans pulled the calf to an opening, and then one of the group rushed back to where the trail from the cave joined the river and retrieved a rock flake that could be used for removing pieces of meat. One of the men set to work cutting while the rest of the group remained watchful. They knew that speed was of the essence to avoid being discovered by hyenas.

Fifteen minutes later, most of the meat from the hindquarters and front shoulders had been crudely separated from the carcass. A man and woman grabbed what remained of the calf and half-dragged and half-carried it upstream beyond where it had been killed by the bear-otter, hoping that its continued presence wouldn't draw the attention of dangerous predators to their dwelling. Two adults carried the meat chunks downstream to the junction with the trail ascending to their cave. The two children carried their spears, and the group moved downstream, constantly watching their back trail.

After abandoning the carcass, the man and woman began retracing their steps to rejoin the rest of the group. The jackals and vultures immediately descended on the bones and viscera, fighting with each other over the choicest morsels. The air was filled with the sounds of their squabbling and the smell of death.

Although pleased with their theft of the meat from the bear-otter, the adults were nervous about drawing attention to their current location. After everyone was satiated from eating what they could, several gathered what remained and the entire group hiked to another cave a kilometer away. More tube than cave, the small cavern had a small opening that only a child could squeeze through that extended back from the surface about three meters. It was a perfect location for keeping meat cool and safe for several days while reducing the risk of predators finding their primary dwelling.

The eleven-year-old girl squeezed down the lava tube with the meat, making two trips. Finally, after all the meat was stashed, the group piled brush in front of the cave-tube entrance and hiked down to the river where they washed away the blood and other smells from the pilfered eland calf. Now late afternoon, they headed back upriver toward their cave. They stopped at a patch of tree heath where they selected and then broke off a large woody stem. After arriving back at the junction to the trail leading to their cave, the woman whose spear was busted in the confrontation with the bear-otter used the cached rock flake to remove all the branching twigs and whittle the end of the tree heath branch into a new spear.

Prior to heading back up the trail to their cave, the entire group drank one last time from the river and relieved themselves. The sunset was a magnificent spectacle of fire and gold, as the sun dipped below the horizon in a blaze of glory. The land was bathed in a warm, golden light. The tall grasses swayed in the gentle breeze, and the air was filled with the sounds of frogs and cicadas in the timeless evening chorus. The day's labors were done, and the darkness crept in, but it was a peaceful end to an eventful day. The group lingered at the mouth of their cave, watching bats flit about, preying on the abundant flying insects in the last dim light. Finally, leaving one of the group at the mouth of the cave to keep watch, the group of hominids retired to sleep and rest—another day complete.

CHAPTER 1

HISTORY

The die was cast for our eventual departure from the planet several million years ago; we cannot justifiably scapegoat a specific generation, political party, country, or social segment for the planet's woes. We've now reached the palliative care phase of our existence on Earth. The conditions occasioning our permanent disappearance are a shared culpability by all humans that have existed for tens of thousands of years—as well as the person we each see in the mirror.

SPECIES

Humans evolved in a world of rich biological diversity. Our ancestors' survival depended upon their ability to identify the plants and animals that provided food, medicine, clothing, tools, and other essential needs and those that were toxic and dangerous. Knowing when fruits and tubers were nutritious and where to find them could stave off hunger for weeks at a time. Avoiding dangerous animals and where they resided meant living another day. Our predecessors developed a mastery in which animals could be hunted for food and where and when they were vulnerable. The ability to clearly convey this information to other members of their tribe was critical. Misunderstandings could easily be fatal.

Humans have gradually developed an increasingly complex system for ascertaining the differences between complex organisms. We have expanded a system whose original sole purpose was survival to being able to distinguish subtle differences among plants and animals, even tracing similarities and differences in the DNA sequence of genes back along an animal's or plant's evolutionary history.

The most referenced portion of this complex scheme is what scientists refer to as a species. A species is a group of organisms having many characteristics in common and that can produce fertile offspring. For example, all humans are members of one species. We may be tall, short, slender, heavy, black, brown, or white, but any male and female human can normally produce fertile offspring.

The distinction between species sharing many similarities can be illustrated by examining the hybrid offspring of the northern spotted owl and the barred owl. In western North America, northern spotted owl populations have been in a steep decline because of logging and competition from the barred owl. When barred owls first invaded old-growth forests in the Pacific Northwest and California where northern spotted owls are native, they would occasionally interbreed. Biologists feared northern spotted owls would gradually be hybridized out of existence by breeding with the bigger and more aggressive barred owls.

Like most owls, both barred and spotted owls hoot to attract a mate or advertise their territory in late winter. The hooting of these two species is distinctive. A series of four hooted notes, with the middle two closest together, is the northern spotted owl's call for marking its territory. The barred owl makes a distinctive call sounding somewhat like "Who cooks for you all? Who cooks for you?" The vocalizations of their hybrid offspring sound like neither barred nor northern spotted owls.

The consequence is the hybridized offspring can't successfully find a mate or nest; they also aren't fertile, and further hybridization didn't occur. Barred and northern spotted owls are, therefore, still two distinct species, moving along separate evolutionary paths.

The value of species recognition can be exemplified by the South American night monkeys that were once regarded as a single species. Scientists relied on them as a laboratory animal for malaria studies. Then a primatologist found there were nine different species of night monkeys that differed in their susceptibility to malaria. The discovery revealed human lives were imperiled by inadvertently testing malaria treatments on species that might not be vulnerable to the disease in the first place. In more sophisticated ways, people's survival is still contingent on knowing and differentiating animals and plants.

Thanks in large part to this widely accepted system, when people talk about northern spotted owls or large-mouthed bass or maple trees or tigers, others clearly understand. At the most basic level, our system of differentiating species is still a mechanism to communicate with each other about other organisms on the planet.

SPECIES EXTINCTION

Extinction in biology is the dying out or extermination of a species. Throughout history, many factors have resulted in species disappearing. Life on Earth is ancient, extending back over 3 billion years. Complex multicellular organisms—what we modern humans think of as life—have existed for more than 600 million years.

Scientists generally agree that Earth is currently experiencing its sixth mass extinction. Five times previously, our planet has undergone changes so dramatic that the diversity of life collapsed. The first occurred about 375 million years ago. Each extinction event resulted in the disappearance from the fossil record of 75 percent or more of known multicellular species at the time. Ironically, the history of these events is being recovered just as people are realizing their primary role in causing another mass extinction.

During the periods between the five previous extinction events, evolution proceeded in a manner that resulted in millions of years of relative ecological stability. The most recent extinction event occurred 66 million years ago and resulted in the disappearance of most of the dinosaurs and other life-forms from the planet. Two theories for their demise currently have traction with paleontologists.

Some scientists maintain that ancient lava flows resulting from a series of eruptions in a region known as the Deccan Traps of current India were the primary cause for the last extinction. These eruptions appear to have occurred over a period of 30,000 years. The resulting volcanic rock from this lava flow is estimated to have covered an area approximately half the size of modern India. The current-day Deccan Traps region has been reduced by erosion and plate tectonics to nearly 200,000 square miles in layers that are in places 6,000 feet thick. Such a vast eruptive event would have choked the skies with sulfur dioxide and other gases and dramatically changed the planet's climate. An average drop of temperature of about 3.6 degrees Fahrenheit (2 degrees Celsius) is likely to have occurred during this period.

Other scientists point to a massive comet or asteroid six to nine miles wide as the culprit. It struck what is now the Yucatán Peninsula in Mexico near the town of Chicxulub. The energy released from the Chicxulub impactor was approximately 100 million times more powerful than the hydrogen bomb, the most powerful human-made explosive ever detonated. The collision between Earth and the asteroid created winds in excess of 600 miles per hour (1,000 kilometers per hour) near the blast's center. With an impact powerful enough to send deadly amounts of vaporized rock and gases into the atmosphere, the effects of this astrological event would have persisted for years.

What seems most plausible is a combination of the two events dealt most dinosaurs and plants a catastrophic scenario. The effect was magnified by each playing out on opposite sides of the planet. One event, or more likely both events, resulted in a mass extinction of three-quarters of the plant and animal species on Earth. It marked the end of the Mesozoic era, ushering in the Cenozoic era that continues today.

The causes of the prior four extinction events remain murky, but there is a continuing theme of either large meteor impact events, massive volcanic eruptions, or a combination of the two. The planet doesn't easily give up its secrets from millions of years ago. There is little evidence of evolution itself precipitating an extinction event until now.

THE SIXTH EXTINCTION

There is ample speculation on how and why we gained the means, motive, and opportunity to perpetrate the most heinous of all crimes against other species: ecocide. Paleontologists are uncovering new information every year that is shedding additional light on a topic that boasts no widespread agreement.

Anatomical Changes

Three separate events occurred nearly a million years apart that may have precipitated the rapidly worsening sixth extinction on Earth. The evolution of the unique structure of the human shoulder girdle was the first and most important and provided the springboard for nearly everything that has since ensued.

It seems increasingly likely *Homo sapiens* is a descendant of *Australopithecus*. These ancient ancestors of modern humans existed in Africa from about 1.9 to 4.2 million years ago. Credit them for space exploration and modern medicine, and blame them for climate change, nuclear waste, deforestation, mass species extinctions, and the host of other insults we are inflicting on the planet.

During their 2-million-year tenure on Earth, the hominid's shoulder evolved to an intermediary between African apes and modern humans. These changes in the anatomy of the shoulder were probably initially driven by the use of tools. The positioning of the shoulder blade with a downward scapular tilt allowed them to engage in both tree climbing and wielding stone tools.

It also eventually spawned our ability to throw. We developed the uniquely human ability to hurl objects with speed and accuracy. The positioning of our shoulder blade allows humans to store energy in the shoulders, much like a slingshot, facilitating high-speed throwing.

Archaeological evidence shows wooden spears were used at least 500,000 years ago and may have been used as far back as 5 million years. They were likely utilized primarily for defensive purposes and as walking sticks. Initially, hunting was limited to occasionally skewering small mammals, ground-dwelling birds, or larger animals already incapacitated in some manner. Any humanoid over the age of five or six probably carried a spear with them any time they weren't sheltering.

Neanderthals were constructing stone spear heads as early as 300,000 years ago. By 250,000 years ago, they were being made with fire-hardened points. About 200,000 years ago, humans began to make complex stone blades with flaked edges, which were used as spear heads.

A recent study found that by 125,000 years ago, the average African mammal was already 50 percent smaller than those on other continents. The advent of complex stone blades appears complicit with some of the latter species extinctions in Africa.

The earliest spears were handheld and consequently short-range weapons. The subsequent appearance of distance weapons about 50,000 years ago represents another consequential evolutionary change in human ecology and social behavior. Widespread use of spear projectiles in a few cases directly resulted in a species extinction but more likely amplified the increasing number of species extinctions, caused by a collection of reasons, in the Late Pleistocene, 10–50,000 years ago in Afro-Eurasia, Australia, and the Americas. These included woolly mammoths that were slightly heavier than today's African elephants, mastodons, elephant-sized ground sloths, and giant beaver.

There are no known complex organisms from other epochs that could throw long-range weapons they had fashioned. It allowed our ancestors to kill animals without touching them or they themselves being touched. To kill something, other organisms must catch them with their feet, claws, or teeth. Spiders could be considered an exception because of catching prey with webs. However, they still must kill their prey manually. The necessity of dispatching prey at close quarters levels the predator-prey playing field. The inability of any other known organism to throw was likely pivotal in providing relatively stable ecosystems for millions of years, before humans developed the throwing ability, only interrupted by geological or astrological events.

While at first it was throwing rocks and spears, over the millennia we've gradually learned how to separate ourselves further and further from both predators and prey. The first improvement was the atlatl, or woomera, an implement allowing more energy and greater velocity to be applied to the throwing of a spear. The throwing arm together with the atlatl acts to increase the length of a lever.

The premise is the same with fishing poles. Although challenging, it is possible to catch fish simply by hand-throwing a fishing line with a lure or bait. Native fishermen have been doing it for decades. However, much more line speed and distance can be attained using a fishing rod because it increases the length of the lever. Or look no further than your local park: ball throwers with those relatively new but already ubiquitous plastic grabbing sticks use the exact same principle for tossing balls for dogs to retrieve. The atlatl is believed to have been in use by early hominids in some parts of the world 30,000 years ago.

The next significant step in weaponry was the bow and arrow. The oldest known evidence of arrows comes from the Sibudu Cave in South Africa. The bone-and-stone arrowheads found there are approximately 60,000–70,000 years old. Despite the early time frame, archery doesn't appear to have been in widespread use until after the planet had lost most of its large mammals. The bow was an important weapon for both hunting and warfare from about 8,000 years ago to the mid-17th century. Bow-and-arrow use was almost certainly contributory to the uptick in animal and bird extinctions during that time frame. More recently, guns, bombs, and weaponized drones kill with limited or no interaction between the aggressor and victim. Humans have essentially turned killing into an enhanced video game.

Domestication of Fire

Concurrent with the advent of complex stone blades and our ability to kill animals at a distance was our ability to use fire. Some human groups may have made occasional use of fire as early as 1.8 million years ago. This may have been the springboard for the movement of early humans into Europe where temperatures are colder than Africa and Asia. By about 300,000 years ago, our predecessors were using fire daily.

Warmth, light, and protection from predators as a result of the domestication of fire were huge factors in our ancestors being able to survive more successfully and utilize areas that were once verboten. Not long after first learning how to use fire, humans may have started deliberately modifying habitats to their advantage by burning them.

Fire allowed humans to cook. It changed the chemistry and biology of foods, killed germs and parasites, and greatly reduced the time necessary to eat and digest food. This decreased human's energy consumption, which may have also helped pave the way for ever-larger brains.

Fire opened a chasm between man and other animals. Prior to fire, the ecological niche animals evolved to fill depended on their bodies: size, strength, speed, claws, teeth, and wings. When humans gained control of fire, they harnessed a potentially limitless force. The use of fire was independent of the structure or strength of the human body. A single human could burn a grassland or an entire forest in a short time.

Behavioral Changes

The third significant development that ratcheted up human domination of the planet occurred 30–50,000 years ago when early humanoids began creating myths. The first evidence of this phenomenon was the lion-man figurine created by Neanderthals and found by archaeologists in a cave in Germany. The body is a human, and the head is a lion. This represents one of the first indisputable examples of art, probably religion, and the uniquely human ability to imagine things that don't exist.

Legends, myths, gods, and religion appeared during this time frame. Humans could communicate with each other about beings and entities they had never seen, touched, or smelled. Fiction enabled people to imagine things collectively and gave them the unprecedented ability to cooperate flexibly in large numbers. Humans could work together with countless numbers of strangers by believing in common myths.

Our ancestors in Eurasia, at least in a few events, interbred with Neanderthals. Their brains were just as big as ours, but ours were likely better at building social bonds. A gene mutation in *Homo sapiens* differs from the Neanderthal version by only one of its amino acid building blocks. While theories as to the effect of the mutation are speculative, the most logical is it made *Homo sapiens* better at building social bonds, imagining things that don't exist, and eventually creating the modern-day extrovert.

This concept has been refined and expanded in modern society. As an example, money is just paper and metal, and has value only because all people believe the same shared myth. Goods and services can be cooperatively bought and sold all over the planet by others who we neither know nor understand but who also believe our money has value. The myths that allowed the cooperation of ancient hunter-gatherer bands may have been shared stories, rather than the financial, social, or religious myths common in modern-day societies.

The confluence of the ability to throw, the use of fire, and the advent of myths contributed to the loss of many of the large mammals in the Americas, Eurasia, and Australia. The mammoth, mastodon, and woolly rhinoceros disappeared. Saber-toothed cats, after flourishing for 30 million years, vanished, as did native American horses, camels, and dire wolves. They were but a few of the many animals that succumbed to a human presence on the landscape.

THE SIXTH EXTINCTION: THE BEGINNING

For the last 2.6 million years, the planet has experienced severe climatic fluctuations during which the Earth waxed and waned between frigid ice ages and warm interglacial periods. Until approximately 60,000 years ago, there were few apparent extinctions. Suddenly, hundreds of species of mammals, flightless birds, and reptiles went extinct largely over the course of 50,000 years. These extinctions marked the beginning of a tidal wave of disappearing species that is cresting today. Because large body size is correlated with slow breeding, the largest animals are more susceptible to extinction.

Even though Africa experienced the earliest large mammal extinctions, the continent escaped with the least amount of damage. Africa is the only continent where large animals and humanoids evolved together. Humans slowly became more deadly and sophisticated at the same time African mammals were gradually adopting behavioral countermeasures. Even in Africa, however, the relatively swift behavioral changes in our early ancestors outstripped the capabilities of many large mammals to adapt.

Most researchers believe that instead of actively hunting large game, australopithecines, our ancient African ancestors, likely consumed whatever edible portions were left on the carcasses of large animals killed by the carnivores that existed at the time. These included the bear-otters described in the prologue, as well as wild dogs, hyenas, bears, leopards, tigers, and lions.

Early in our history, what remained after the carnivores were satiated was probably bones. It is widely believed humans broke the bones and ate the bone marrow, the calorie-rich gloppy, spongy substance inside bone. Because the marrow is encased inside bones, they were much less apt to acquire harmful bacteria than by eating residual muscle tissue that was exposed to warm temperatures for an extended period.

Scavenging bones from under the noses of big cats, wolves, and other large predators was a risky business. Like the vultures and marabou storks of Africa, our human ancestors likely waited in the safety of trees or at a safe distance for the dangerous predators to finish and leave before venturing to a carcass.

Continual monitoring of the sky for vultures and other carrion-eating birds was our ancestors means for finding fresh kills. Some things haven't changed in 4 million years. Whenever in the field to find or document carnivores, I was constantly looking for tracks and listening and scanning the skies for scavenging birds. In wild places on all continents, there are a contingent of avian scavengers including some combination of eagles, vultures, buzzards, ravens, and kites.

While tracks are an excellent way of documenting where a large carnivore has recently been, finding birds concentrated on a kill was often a means of knowing their current location. Some opportunistic birds who scavenge carcasses often follow large carnivores around knowing sooner or later they will kill an animal and food will become available.

Until the Agricultural Revolution, about 12,000 years ago, climbing trees or hiking to locations with a view to continually monitor the skies for birds was a necessary activity early humanoids performed on a daily basis. Monitoring the location of large predators by watching and listening for birds to give them advanced warning of their presence was an obligatory survival strategy.

Our ancestors became adept at assessing risk and reward when scavenging the carcasses of dead animals as a result of several million years of trial and error. Over the millennia, they gradually learned how to work together more efficiently in driving off carnivores to minimize the risk to individuals. At the same time, their weaponry was becoming increasingly sophisticated.

This resulted in being able to insert themselves into the process of scavenging the edible portions of a dead animal earlier. Gradually, between approximately 250,000 and 2 million years ago, rather than skirmishes with jackals and vultures over what remained of a carcass, they could drive off ever-larger predators, occasionally even solitary large carnivores who had actually done the killing. This may have been the single most important development precipitating the extinction of megafauna.

There is safety in numbers. This hasn't changed in the interactions between large predators and humans in the last 2 million years. During the daytime, 10 to 15 humans armed with spears were increasingly able to drive off a single lion, bear, hyena, or wild dog. One or two humans trying to do the same would have resulted in them being converted into tasty hors d'oeuvres by the large carnivore.

Safety in numbers is still true today. While extremely rare in modern times, people killed by grizzly bears and mountain lions are almost always by themselves. Occasionally there are two people present, and rarely three. I'm unaware of any attacks on a group of people larger than three.

Our ancestors precipitated the extinction of both large carnivores and plant-eating animals in many cases without killing either. All that was necessary was continually monitoring the skies and when they determined a kill had been made, drive the carnivore off the carcass as soon as possible. The carnivore was then forced to either kill another large plant-eating animal, also called herbivore, or starve.

The inevitable outcome over time was the extinction of both. Even if they were successful in acquiring another animal before starving, the fragile predator-prey balance that had evolved over tens of millions of years was destroyed. The increased predation resulted in both the predators and their prey becoming increasingly vulnerable to extinction from rapidly declining numbers. Once the large animals were few in number, inbreeding, changing climate, or direct attacks by humans would have sealed their fate.

Trophic cascade is the term used for this type of ecological effect by a single species. Trophic cascades are typically caused by the loss or introduction of a single major predator from or into an ecosystem. With the elimination of wolves in the American West, and particularly Yellowstone National Park, elk numbers skyrocketed. Elk overgrazed streams, meadows, and aspen forests negatively affecting many species of birds, insects, fish, and mammals. Even after only 25 years with wolves back on the landscape in Yellowstone, these processes have begun to reverse themselves.

The calamitous cascade of species declines currently being experienced by ocean megafauna due to human harvesting mirrors what has happened on land. Although there has been little absolute extinction, there has been a dramatic decline in the abundance of whales and large fish through over-harvesting. Both on land and in oceans, declines currently being documented are but an extension of what transpired during the Late Pleistocene with the large terrestrial mammals.

Many of these large animals were present in low densities, were long-lived, and had low reproductive rates. The herbivores were naturally regulated by predators like saber-toothed cats, dire wolves, and lions. Because humans were also omnivores, they could switch their diet to plant foods and smaller mammals when populations of the large herbivores were low or collapsed. Without the flexibility to switch to other foods, the remaining large predators were steered into extinction.

Humans' inability in some instances to quickly insert themselves between predators and their prey may partially explain why Africa and Asia didn't experience the same level of extinction of megafauna that Australia, Europe, and the Americas did. Today, East Africa has six carnivores that are large enough to be considered megafauna. These include lions, leopards, cheetah, and several species of hyena. Once upon a time, as many as 18 large carnivore species shared the East African landscape including omnivorous bears and civets, saber-toothed cats that specialized on large prey, and the bear-sized otters described in the prologue.

A closer look at the ecology of the extinct carnivores, and more importantly those that remain, may help explain human impacts on megafauna. Even armed with spears, 10 or 15 hunter-gatherers were not going to drive a pride of lions off a kill. Prides of lions vary in number between eight and 40 individuals. In a human attack, the lions might sustain some casualties, but the hunter-gatherers would have been annihilated. For that reason, this is a scenario that early hominids probably avoided at all costs. Safety in numbers worked just as well for the lions as it did for humans.

To escape hyenas and lions, leopards drag their kills into trees or caves. It would have been equally effective in escaping humans. A mass attack by humans on a leopard perched on a tree limb or in a cave wasn't an option. The result of dragging their kill into tree canopy and caves also made it less obvious to birds flying overhead and avoided the widespread telegraphing to humans inhabiting the area. Hyenas also utilized mass tactics by operating in larger groups. Cheetahs usually kill small antelope that wouldn't have been worth the metabolic effort of an entire group of hunter-gatherers to scavenge unless it was coincidental. The unique strategies of lions and leopards of either operating in large groups or dragging their kill into trees and caves saved them and many of their prey species from the reign of extinction wrought by humans in Africa.

When humans migrated out of Africa where animals had never experienced human predators, extinction rates increased dramatically. The increased rate of large animals permanently disappearing across the planet mirrored the sequential pattern of the migration of anatomically modern humans.

Hunting is defined as pursuing, chasing, or ambushing a wild animal for sport or food. Walking up to an unsuspecting megafaunal animal and killing them with spears or clubs without ever being recognized as a predator is not hunting; it's an execution. Most megafaunal predators in Australia were amphibians and reptiles. Humans neither looked nor acted like these animals and went unrecognized as dangerous by the large birds and animals. Many megafaunal animals across the planet were slow-moving and had evolved no defenses against a spear-throwing or club-yielding bipedal human. The primary defense of giant grounds sloths, moas, and nearly the entire complex of megafauna in Australia against predation was their size. They were quickly lost in the execution-style killings by the first wave of humans into new areas.

There were many variables in play with how quickly megafauna extinction time frames occurred following a human presence on the landscape. The extinction could be sudden or gradual. It could happen at first contact with humans or after millennia of interaction. Terrain, climate, weather, adaptability, and animal behavior are but a few of the pertinent factors. In some cases, an animal's vulnerabilities may have been quickly exposed. In others, it may have taken centuries for humans to discover their susceptibility or for a natural event to occur that temporarily rendered them in great peril. As already described, extinctions may have been precipitated by humans but only indirectly.

In circumstances where wild animals rarely, if ever, encounter humans or the experience doesn't result in injury or death, they often don't equate human presence as a danger. This should come as no surprise to those familiar with hunting. Many big-game animals that experience intense hunting pressure learn better strategies for avoiding humans. For those animals that have behaviorally adapted to living in or near urban areas where hunting is prohibited, they no longer equate humans as a danger. These behavioral changes can happen in just a few decades.

Another life history trait of the large ancient carnivores that would have been a vulnerability when humans arrived on the landscape was denning sites, either for hibernation or for giving birth and raising young. Because of the cooler climate in much of Europe and North and South America, caves may have been more important for the survival of carnivores in these areas than those living in more temperate areas like the African savanna and Asia.

Many of these caves were probably the same ones coveted by our early ancestors. As the result of our predecessor's ability to use fire, carnivores would have been vulnerable to being killed while hibernating, driven out, or excluded from their dens when discovered by humans. Even a large group of social carnivores would have been in jeopardy. Once the animals exited the cave, humans could easily move in and block their return with a fire at the entrance. Little is known about the ecology of extinct carnivores, but if they were limited by available denning and hibernation sites, the loss of these caves to hominids may have been enough all by itself to precipitate a downward population spiral and doom them to extinction.

Denning sites utilized by carnivores still in existence today are fairly subtle, obscure, and difficult for humans to find. They aren't the kinds of caves or shelters where humans themselves can live or seek shelter. It might be the primary reason why grizzly and black bears, wolves, and cougars still persist in North America when their larger cousins were driven to extinction.

In North America, there is documented evidence of humans working in larger groups driving and then trapping herds of animals in canyons or pits where they could be easily killed with spears thrown from the safety of higher vantage points. Horses, camels, and mammoths would have been susceptible to this hunting tactic.

These same herds of horses and camels would also have been a frequently targeted prey of the larger and more social carnivores that existed on the continent. With the occasional elimination of herds of these animals by humans, it may have directly driven social carnivores to extinction, forced them to become solitary rather than group hunters, or both.

For millennia, there has been a weather event that temporarily renders many North American herbivores extremely vulnerable to predators. In the event of a heavy snowfall, followed by a brief warming period and then another sudden cold snap, the snow mantle freezes. At that point, it becomes a simple exercise in physics. Animals with a small foot or hoof relative to their body mass break through the snow. Those animals with larger feet or smaller body size don't and can walk on top of the snow mantle, expending little energy.

Animals like wild horses and camels that were relatively invulnerable most of the time would have become easy prey for early human hunters during these weather events. While the native carnivores only killed what they could immediately utilize, it is extremely likely our ancestors killed everything they could when this occurred. More details regarding this phenomenon are provided in chapter four.

There are always cascading effects that result from the disappearance of the largest animals on a landscape. The loss of megafauna in the Pleistocene resulted in the extinction of many plants and smaller animals. It would have also affected the relative abundance of many others. The largest animals always have disproportionate effects on natural ecosystems.

Some factors are simply unanswerable with research—weather, hunting efficiency and dynamics, and changes in human and animal behavior. Only in controlled laboratories do things happen in a vacuum. The real world is much messier, with chaos and chance often reigning supreme. Changing climate may be interwoven with these other factors as a contributory cause of extinctions resulting either directly or indirectly from trophic cascades and overhunting.

Climate and food availability have affected species throughout history and resulted in population fluctuations long before humans arrived on the scene. For several million years, Earth has gone through cycles of warming and cooling that resulted in glaciers advancing and then receding. However, mass extinction events have never been documented as a result of these gradual climatic changes. There is no compelling evidence that the last glacial-interglacial change was so different from previous ones that whole-scale extinction of the largest mammalian species on the landscape would result. Those species had survived previous glacial-interglacial transitions.

Species incapable of long-distance movements can be affected much more profoundly by climate change than most megafauna. Their inability to move makes plants, small vertebrates, and invertebrates more susceptible to the effects of climate change than larger animals with greater mobility. However, because these more obscure species are the foundation of the food chain, the elimination of these species can also have a bottom-up effect on larger species, especially when they are present in great abundance. The Rocky Mountain locust is an example of this, whose disappearance is described in chapter three.

In some cases, humans were documented as being present on the landscape for several thousand years before the extinction of an animal occurred. For this reason, some theorize that climate change is the obvious culprit. Humans were in North America for nearly 20,000 years before the emergence of the first Apple computer or the disappearance of the California grizzly, but most people agree a changing climate wasn't a factor with either. It's easy to lampoon theories, but in their defense, deciphering causes of a species' demise millennia after the fact is an incredibly difficult, complex, and sometimes impossible undertaking.

What seems clear is that megafauna in the Americas and Australia were annihilated. Australia, North America, and South America, which respectively had the highest incremental extinction rates, had no known native species of apes at all, and specifically no species of humanoids. The farther the distance from Africa, the greater the losses incurred by megafauna.

However many of these factors are intertwined, the megafaunal extinctions associated with the Late Pleistocene correlate with the relatively short time between when humans first appeared and when the species vanished forever. The beginning of the sixth extinction was a confluence of enhanced tool-making abilities, improved throwing skills, domestication of fire, migration into new geographic regions, learning and exploiting the vulnerabilities of animals in those areas, greater flexibility by early humans in working together in larger numbers, and natural events occurring simultaneously. The first wave of humans was one of the biggest and swiftest ecological disasters to befall the animal kingdom.

Individual men and women in hunter-gatherer societies immediately prior to the Agricultural Revolution represent the pinnacle of human evolutionary development. Their athleticism was similar to modern-day decathletes with one notable difference. Instead of retiring when they were 30 to 35 years old, as do contemporary professional athletes, hunter-gatherers maintained most of their athleticism into their forties, fifties, and sixties. Their average longevity was significantly shorter than current-day people because of a high rate of infant mortality. Those who survived the hurdle of infancy, however, often lived well into their sixties and even seventies.

For the last 2,000 years, infant mortality has hovered around 27 percent, and about 46 percent of youths died before reaching puberty. Prior to the Agricultural Revolution, infant and youth mortality was almost certainly even higher. Only in the last 150 years has infant/child mortality waned significantly.

For millions of years, childbirth may have been the leading cause of hominid mortality, for both newborns and pregnant women. Today, giving birth is 300 times safer than it was even a few generations ago.

The maternal death rate during childbirth in the 1800s was estimated at 5–10 per 1,000 live births. Historically, being a fertile human female was a dangerous business.

Accidents, disease, and predation were factors that exacted the heaviest child mortalities prior to puberty. The skull of the Taung child (*Australopithecus africanus*), one of the most studied skulls ever and long presumed to be killed by a predatory cat, was more recently identified as having been prey of a large raptor. Birds of prey are often overlooked as potentially having been a significant predator of hominid young. Children between the ages of one and four would have been susceptible to predation by raptors in Africa, particularly the crowned hawk-eagle. These birds have been recorded preying successfully on adolescent mandrills and young bonobos that weigh 20–50 pounds. Antelope weighing around 40 pounds are routinely killed by these powerful raptors. Eagles typically dismember their prey quickly, so load-lifting capacity is not an issue. The juveniles of all large and medium-sized carnivores and herbivores are potential prey of large raptors. In addition to birds of prey, early humans were hunted by wild cats, bears, hyenas, snakes, and crocodilians, among others. Adolescents would have been the most susceptible to being preyed upon.

Modern humans have no appreciation for the incredible amount of knowledge our hunter-gatherer ancestors retained to survive in what was often hostile environments. Within the area they wandered, every individual likely knew all the different birds, their calls, and when and where they nested. They would have known when fish spawned, where reptiles and amphibians congregated, and what insects were palatable. They knew more about wildlife than any modern-day biologist, more about plants than any botanist, and more about the uses of plants than herbalists. They almost certainly understood basic first aid. They were Siri, Alexa, Google Earth, a topographic map, and a global positioning system unit all rolled up into one individual.

They knew where to acquire basalt and flint in order to flake points, needles, and knife blades and had the fine-motor skills necessary to do so quickly. Their powers of observation and ability to use not only their eyes but their hearing, smell, touch, and taste far exceed our own. They could endure temperature swings that would leave most modern humans either hypo- or hyperthermic.

They had mastered the internal world of their own bodies and senses. They listened to all the bird and animal sounds that advertise the presence of a lurking dangerous predator. They carefully observed the foliage of trees to discover fruits, beehives, and bird nests. They moved with minimum effort and noise and knew how to sit, walk, and run in the most agile and efficient manner. The inability to have all these skills and knowledge at their disposal usually meant them being quickly swept out of the gene pool. Nature rarely gave our early ancestors a second chance to make a first impression.

Their sense of self-preservation was finely honed. COVID fatigue has become a common phenomenon among many individuals in contemporary society, leading some to overlook the ongoing risks posed by the virus, even shortly after the pandemic began. In contrast, early *Homo sapiens* never experienced the luxury of becoming fatigued by the dangers of saber-toothed tigers or poisonous plants. Those rare few who did succumb to such fatigue were swiftly eliminated from the population.

Their understanding of personal place among their animal and plant neighbors and the terrain, was just as acute. One of my colleagues spent more time in the field than any other I knew. Working out of the same office in the same area his entire career allowed him to become intimately familiar with the terrain, plants, and wildlife. He knew nearly every mountain lion in a 1,000-square-mile area, where they crossed roads and streams, and how frequently they moved through an area. He spent about 300 days in the field every year.

By contrast, hunter-gatherers spent 365 days a year, 24 hours a day in the field. They knew the haunts of nearly every large predator and herbivore within the area they roamed. Rarely did they not know the risk of death when hunting or gathering in any given area. Risk assessment was a natural part of their every activity. Their grocery store was an ever interesting, rewarding, and dangerous place to shop. People equate our opulent lifestyle as proof of supreme intelligence. But our hunter-gatherer ancestors were tougher, more athletic, and more knowledgeable than we are today. This phenomenon is discussed in detail in the following chapter.

LEOPARDS

Humans and leopards have had a disharmonious marriage transcending human evolutionary history. The historical geographic range of leopards (*Panthera pardus*) and humans exhibit an astonishing level of overlap, encompassing regions where ancient hominids thrived. This extensive convergence of human and leopard habitats fostered a profound connection between our ancestors and these elusive predators. Among all mammals, it is unlikely that any other species has exerted a more profound impact on our development and who we have become today.

Leopards are renowned as solitary and stealthy hunters that are powerful runners, swimmers, and climbers. The only animal to come close to matching them is the African lion. Lions seize the chance to kill leopards whenever it arises, but leopards often manage to find refuge in trees, ensuring their safe escape. The presence of trees or rocks as refuge provides leopards with the ability to coexist alongside their larger rivals. While lions are capable of maintaining exclusive territories devoid of other lions, their inability to exclude the agile and tree-dwelling leopards is a testament to the latter's nimbleness and adaptability.

Leopards once roamed across vast regions in Eastern and Southern Asia, as well as the entirety of Africa. Presently, their distribution is primarily confined to sub-Saharan Africa, although remnants of populations persist in Indonesia, Pakistan, India, Sri Lanka, Indochina, Malaysia, China, North Korea, and eastern Russia. As is the case with most native animals, leopards have experienced a significant decline, disappearing from nearly

40 percent of their historic African range. Despite their remarkable adaptability to diverse environments, including those shared with humans, their elusive nature remains unchanged. Even in proximity to human settlements, they continue to lead secretive lives.

Like all mammals, humans have fundamental requirements for survival, including access to vital elements such as food, water, and secure shelter. The formidable threat posed by leopards and African lions to our ancient ancestors played a pivotal role in regulating their population dynamics and presence on the landscape because safety always had to factor into decision-making. Some regions offered abundant food and water resources yet lacked safety, while others provided safety but suffered from limited access to sufficient water resources.

Throughout human evolutionary history, leopards and African lions were likely the predators exacting the greatest toll on adults. There are still occasional reports of these two carnivores preying on humans, including rare instances of tourists falling victim to lion attacks. Two notable cases from the late 19th and early 20th centuries gained widespread attention and were romanticized in books and movies. The Leopard of Rudraprayag in India, responsible for the deaths of over 125 people, and a pair of African lions from Tsavo, which caused significant disruption during the construction of a railroad in 1898, resulting in the deaths of dozens of workers, remain enduring examples.

Among these two carnivores, leopards undeniably posed the greater peril. The size of lion prides often rendered their presence noticeable to other African wildlife, enabling people to be alerted to lions through their observations of the sounds and behaviors of birds and other animals. Climbing trees served as a largely effective means for humans to evade African lions. However, the most significant danger posed by lions likely stemmed from dispersing young adults in search of their own territories and mates. Traveling individually, their reduced detectability by other wildlife and the unpredictability of their movements would have presented a serious threat to ancient hominids.

The strategy of seeking refuge in trees proved futile for ancient hominids when confronted with leopards. The animals' solitary and nocturnal nature, combined with their exceptional tree-climbing abilities, meant that encounters with these predators often happened suddenly, without warning or the possibility of finding refuge. Leopards posed a formidable threat, and they emerged as our ultimate adversary.

Leopards utilize caves as shelters, feeding sites, and breeding lairs whenever they are available. Studies indicate that when caves are present, leopards stash their prey there; they use trees otherwise. These remarkably adaptable creatures exhibit equal comfort in the illuminated, twilight, and dark areas of a cave, often opting to feed near the entrance. However, they retreat to the darkest interior when necessary, such as during the birthing process, to protect and feed their cubs, or to safeguard a kill from scavengers like vultures, jackals, hyenas, and lions. The cooler temperatures within the caves significantly slow meat spoilage, enabling leopards to store larger prey items. Instances of zebras and elands being cached have been documented. In hot and arid environments, leopards seek refuge in caves to escape scorching daytime temperatures and minimize water loss.

It is unclear whether leopards consciously prioritize larger animals when caves are available for caching prey. When leopards rely on trees as their primary means of avoiding scavengers, they tend to target small-to-medium-sized antelope and similarly sized animals.

Wonderwerk Cave in South Africa is widely acknowledged as one of the earliest documented sites providing evidence of human utilization of caves, dating back over 2 million years. The presence of humans in caves became more prevalent between 2 million and 20,000 years ago, eliciting the modern-day moniker "caveman." It is highly probable that leopards and ancient humans occasionally shared the same cave, whether knowingly or unknowingly.

Baboons are frequently targeted by leopards, particularly as they sleep in trees or caves during the night. Conversely, during the day, baboons turn the table, exhibiting aggressive behavior, actively attacking, displacing, and occasionally killing leopards. This fluctuating day-night dynamic in their predatory-parasitic relationship may have also been mirrored in the interactions between leopards and our early hominid ancestors.

In South Africa's Sterkfontein Valley, a paleontologist discovered fossil remains of hominids, baboons, antelopes, as well as leopards and other large carnivores within the ancient vertical shaft caves. Analysis of the prey animal sizes, the specific body parts recovered, and the puncture marks on some of the cranial bones of hominids and baboons strongly suggests that many of these fossils represent remnants of meals consumed by leopards. It is plausible that these bones originated from trees where leopards fed, which grew out of the cave entrances, and the bones subsequently fell into the caves.

While leopards were predators of our early ancestors, hominids also derived benefits from cohabitating near these carnivores. Leopards often abandon their kills for periods ranging from three hours to two days, leaving behind substantial amounts of flesh and nutrient-rich marrow. By scavenging from these leopard "larders," early hominids were able to obtain the meat and marrow from small-to-medium-sized prey animals in relative safety. The tree-stored leopard kills likely served as a valuable resource for early scavenging hominids. Despite occasional theft of portions of the kill, leopards, being efficient hunters, were likely able to sustain their success. Hominids may have utilized the sharp, fractured limb bones of partly eaten carcasses to peel back hides, expose flesh, and remove large muscle bundles. This activity might have even served as an initial impetus for early hominids to create and utilize tools in the extraction of animal nutrients.

The intricate and dynamic interaction between humans and leopards likely played a crucial role in preventing hominids from attaining dominance. However, the acquisition of fire disrupted the long-standing equilibrium that had persisted for hundreds of thousands of years. The mastery of fire by humans provided a new level of protection against leopard predation in areas they continued to cohabitate. It also facilitated their expansion into colder climates and effectively transformed humans into an invasive species.

The deliberate targeting of those of lesser power is what distinguishes bullying from garden-variety aggression. It involves acts of aggression and exploitation. For several million years, the reward of aggressive, violent behavior was our survival. It came at the expense of carnivores that co-existed with humans on the landscape. With the abandonment of kills, leopards provided hominids a relatively consistent opportunity for thievery. While bullying and theft were evolutionarily inspired for survival, these behaviors haven't disappeared.

Ours has been a subtle metamorphosis from an animal who scavenged, stole, and gathered to one furthering our economic interests, political ambitions, and social status and imbued with a sense of cultural superiority. We are our ancestors in high heels, suits, and ties.

THE SIXTH EXTINCTION CONTINUES

The extinction of the planet's large mammals is often referred to as the Late Quaternary Extinction Event. This mischaracterization of their demise as an almost mystical, isolated event has created an incredible amount of confusion. The Quaternary Extinction Event was nothing more than the beginning of Earth's sixth mass extinction, which continues today and will end with humans' demise. The last of the megaherbivore extinctions in North America were woolly mammoths, who clung to existence on Wrangel Island until 4,000 years ago. At the time of their extinction, there were civilizations in many parts of the world, including Mesopotamia, ancient Egypt, Minoan, Ancient China, and Norte Chico.

By the beginning of the Agricultural Revolution, *Homo sapiens* had driven to extinction about half of the planet's large animals. Increasing numbers of extinctions of smaller plants and animals have been documented throughout the Agricultural and Industrial Ages.

There are an estimated 15 million different species on Earth. Only 2 million of them are known to science. Because we are unaware of many species on the planet, we are also blissfully oblivious of many extinctions we have already precipitated. It is estimated that 10,000–100,000 multicellular organisms are currently shepherded to their extinction every year by humans. We are losing species a thousand times faster than the normal rate of extinction. A species disappears about every 30 minutes. In some cases, new roads and rapid deforestation are opening habitats once too remote to explore. Researchers sometimes discover new species just as farming, deforestation, and urban development are pushing them to the brink of extinction.

For millions of years, humans scavenged the carcasses of large animals killed by carnivores, hunted smaller animals, and gathered what they could. The previously described factors rocketed us from the middle of the food chain to the top. It turned humans into the most dominant species on Earth. The remarkable ability to throw came with the unfortunate and nasty side effects of millions of rotator cuff injuries annually and our imminent extinction.

The evolutionary adaptation of our shoulder may have been the threshold that began the sixth extinction on the planet. The demise of hundreds of thousands of species became inevitable. Their tipping point is ours, and ours is theirs. A lack of extinction synchronicity doesn't mean they weren't set in motion by the same evolutionary event.

Most people erroneously believe that as a result of the asteroid and volcanoes, one day there were dinosaurs and the next there weren't. The effects of past massive lava flows and meteor impacts on some species was immediate and dramatic. Others clung to existence for hundreds and even thousands of years afterward before finally expiring. It's unlikely the extinction of tyrannosaurs and triceratops occurred in the same decade or even same century. However, whether it was massive lava flows, a meteor impact, or both, the tipping point for all the dinosaurs was the same.

Humans are chronicling an unfolding drama of our own making where we are both protagonists and victims. The script has already resulted in the demise of many species, with many more to follow, including the lead character. The nearly 2-million-year-old sixth extinction is now nearing its conclusion.

CHAPTER 2

INTELLIGENCE

There is only one paradigm that has been adopted universally by the scientific, religious, social, and financial establishments. The most unquestioned assumption in recent human history is that *Homo sapiens* is the most intelligent organism on the planet. It is so widely accepted that it has now become unassailable dogma.

MYTHS

No one needs convincing of our brilliance. All the deities we've created over the years say it's so. In fact, many religions teach that all other organisms were put here for our use and enjoyment. There is no need to be a religious person to believe we are the most intelligent organism. Atheists and agnostics do too. It might be the single point of agreement for people of every race, gender, culture, education, and religion.

Intelligence is an abstract idea created by humans. Each person has their own concept of what intelligence means, including accomplishing complex cognitive feats; achieving goals; resolving genuine problems or difficulties; acquiring, storing, retrieving, combining, comparing and using information and conceptual skills in new contexts; thinking rationally; dealing with their environment; and adapting to circumstances.

All these descriptors can be used in comparing the relative intelligence of humans in the world today. However, people who are considered intelligent today may not have been able to even survive ten thousand years ago and vice versa.

The first attempts to quantify intelligence date back to 1883. They were set up by Englishman Sir Francis Galton, who believed intelligence was largely a product of heredity. He hypothesized that there should exist a correlation between intelligence and observable traits such as reflexes, head size, and muscle grip. The effort was abandoned when he was unable to show any correlation.

Another attempt, in 1905, focused on verbal abilities. The French psychologist Alfred Binet along with Theodore Simon and Victor Henri published the Binet-Simon test with the intent of identifying mental retardation in school children. Binet thought that intelligence was multifaceted but came under the control of practical judgment.

In 1910, Lewis Terman, an American psychologist, revised the Binet-Simon test, which resulted in publication of the Stanford-Binet Intelligence Scale in 1916. It became the most popular test in the United States for decades. During World War I, the US Army tested 1.75 million men, making the results the first mass-produced written tests of intelligence. While the results were dubious, this effort precipitated even larger efforts that gradually evolved into today's intelligence quotient (IQ).

In modern-day society, relative human intelligence is generally thought of in terms of IQ. However, few people know the IQ of friends, family members, and in most cases even themselves. Irrespective of definitions, psychologists, and IQ, nearly everyone makes their own subjective assessment of whether people they interact with are intelligent. It doesn't matter if their IQ is 60 or 160. "My boss is an idiot," "The governor is dumber than a post," "Her friend is a few bricks short of a full load," "Those college-educated kids have no common sense." The quips of people describing others' intelligence go on and on.

The single factor most commonly utilized by individuals assessing other people's intelligence is the degree with which they share myths. The myths humanity has created over time provide the framework of modern society.

They affect every aspect of our lives. Our reality is governed by religious, national, judicial, social, business, and financial myths.

In social circles, they are known as fictions, social constructs, or imagined realities. Belief in some myths is necessary just to survive in today's world. Money, corporations, and justice are all examples. Other myths like religion, human rights, nationalism, and marriage are optional. It's similar to eating at a buffet. We can pick and choose which optional myths we like and which we don't. We can believe the optional myths a little, a lot, or not at all.

While many myths may seem optional, there is a great deal of pressure by society on individuals to conform and accept them. The pressure from friends, family, and colleagues to conform is even greater. A person's happiness, acceptance by society, and self-worth may depend on believing commonly held myths of the family, community, workplace, and nation in which they reside.

Belief in some myths may be inescapable in some geographic regions and optional in others. A recognition and belief in the existence of the homeland or flag may be necessary for survival in some countries. In others it may be convenient and beneficial but not mandatory.

Sharing a single myth is often sufficient for people to immediately believe they are more intelligent than those who don't. A religious person may immediately believe others in their congregation are more intelligent than those who worship a different god. A committed military person may innately believe another military person is more intelligent than a civilian. A millionaire could do the same for their rich colleagues over a poor person.

The perception of intelligence increases exponentially for people who fervently believe and share multiple myths. A rich businessman who believes in white supremacy will likely believe that others who share their wealth and racist beliefs are more intelligent than those who only share their wealth or only share their belief that whites should dominate society. Generally speaking, the more we share myths, the more intelligent we perceive other people with whom we directly or indirectly interact.

To illustrate this on a personal level, several years ago I was invited to a family member's home for dinner. We hadn't interacted closely with each other in many years. They are deeply religious and knew I had grown up going to church with my family and probably assumed nothing had changed. After a nice dinner, the conversation shifted to religion, and they asked me what I believed. When I told them I was an atheist, one of them blurted, "But you seemed so intelligent!"

An additional factor people frequently use to assess other people's intelligence is shared skill sets. Most people regard other people in their profession or who enjoy their hobbies as having more intelligence than the population at large. The rich, racist hypothetical person described earlier with a golfing hobby would probably believe another golfer who shared his beliefs was smarter than someone who watched NASCAR.

The combination of shared myths and skill sets are often the framework for close friendships. Most people gravitate to like-minded people. Nearly everyone thinks their current best friend is intelligent. Likewise, few people think, "I'm not very smart." Occasionally, people say that in a self-deprecating, humorous way, but rarely seriously. Whether it is articulated or not, most people believe they are a stable genius. The flip side is most people perceive others who are different from them as less intelligent. The greater the disparity, the less they regard others as intelligent.

The phenomenon isn't unique to intelligence. It also has been documented with driving skills. Because there is no standard definition for "good driving," people tend to use their own unique, individual definitions. The ability to text while driving might be one of the characteristics that a person believes makes them a unique and superior driver. Even though drivers are aware that others are using different standards for good driving, people often view their own standards as superior. It's easy for a person to be intelligent because, like good driving, they define it.

While all these things are true in how we perceive the intelligence of our fellow humans, we do the same to other organisms with which we share the planet. There is no evidence, thus far, that other species create myths. Ironically, we believe other organisms are incapable of intelligence partly because of their insistence on living in an objective reality. Most of the various religious myths we've created since the beginning of the Agricultural Revolution provide the supportive rationale for dismissal of other organisms as stupid objects to be used and abused as we see fit.

TECHNOLOGY

We're the only organism that has explored space, built hydroelectric dams, and created Ultra HD smart televisions. We're the only species that has developed an electron microscope, submarines, and drones. We're the only species that has created an environmentally friendly bubble around ourselves where we rarely venture into temperatures above 78 degrees Fahrenheit or below 68 degrees or need to encounter rocks, thorns, claws, dirt, or mud. Our rationale for superior intelligence is in part based on these complex wonders we have created.

Our hubris prevents us from seeing this as exactly the same argument creationists use to refute evolution. How could a bear's ability to smell a dead carcass 18 miles away or an eagle having eyesight four to eight times stronger than a human's really be the end product of a long sequence of random but purely natural causes? A giant sequoia is just too complex, too massive, too beautiful to have come into existence except by intelligent design, some people assume. Even evolutionists blind themselves to our own organized complexity spawned from simple beginnings and numerous small steps made by many people whose primary focus is on change, rather than from superior intelligence.

Although Darwinism may not be directly relevant to the world of technology, it raises our consciousness in areas outside its original territory of biology. It teaches us to seek out graded ramps of slowly increasing complexity. If evolution, a nonthinking, random process, can create complex living creatures, why do we think that our superior intelligence is the reason we have today's technology?

Similar to evolution, our most advanced technologies were built on modest beginnings, tinkering and experimentation, and slowly increasing complexity. Most people believe Benjamin Franklin discovered electricity. However, his famous kite experiment only demonstrated that lightning and tiny electric sparks are the same thing.

The first documentation of electricity dates all the way back to 500 BCE when Greeks discovered static electricity by rubbing fur on amber. Two thousand years later, in the 1600s, an English physician and physicist William Gilbert published the first theories about electricity in his book *De Magnete*. The next major text about electricity, *Experiments and Notes about the Mechanical Origin or Production of Electricity*, was published in 1675 by English chemist and physicist Robert William Boyle.

In the early 1700s, decades before Franklin's kite, English scientist Francis Hauksbee made a glass ball that glowed when rubbed while experimenting with electrical attraction and repulsion. The glow was bright enough to read by, and this discovery would eventually lead to neon lighting a few centuries later. In 1800, Italian physicist Alessandro Volta discovered that particular chemical reactions could produce electricity. He constructed the voltaic pile, an early electric battery that produced a steady electric current and in doing so became the first person to create a steady flow of electrical charge.

Michael Faraday created a crude power generator in 1831, which solved the problem of generating electric current in an ongoing and practical way. This opened the door for American Thomas Edison and British scientist Joseph Swan, who in 1878 each independently invented the incandescent filament light bulb. Four years later, they worked together to provide power to illuminate the first New York electric streetlamps.

In the late 1800s and early 1900s, Serbian American engineer Nikola Tesla made important contributions to the design of the modern alternating current electricity supply system. Later, American inventor and industrialist George Westinghouse purchased and developed Tesla's patented motor for generating alternating current and was influential in providing the framework for modern electricity.

This only describes a few of the many individuals involved in making electricity a complex convenience we take for granted. Numerous other scientists, physicists, mathematicians, and inventors contributed important discoveries along the way. Each built on the work of their predecessors in developing our modern electrical grid.

The parallels between evolution and the technologies we currently enjoy doesn't stop with the complexity of the end product. The number of new and complex technologies increase as more people are devoted to research and development. Innovative companies almost always have large research and development departments. Similarly, the speed with which changes manifest themselves through genetic expression can be influenced by either the size of the sexually active population or generation time and how frequently they reproduce.

The pesticide chlordane was first used in homes in 1948. It was thought to be invincible because it was so toxic to insects. By 1951, German cockroaches in Corpus Christi, Texas, were resistant to it. The roaches were 100 times more resistant to the pesticide than laboratory strains of the insect, which normally have less genetic diversity and whose nutrition and disease are effectively managed so they display consistent performance. By 1966, the roaches had also evolved resistance to malathion, diazinon, and fenthion. Soon thereafter, they were discovered to have developed a resistance to DDT. Each time chemists cooked up a new pesticide, it was just a short period of time before some population of the pests developed resistance. The roaches spread and thrived.

Under the stress of intensive chemical spraying, the members of the insect populations with no natural resistance to chemical applications were weeded out. Only the strong and fit remained to defy the attempts to chemically control them. These were the parents of the new generation, which by simple inheritance possessed all the qualities of "toughness" inherent in its forebears. After a few generations of strong and weak insects and continued application of pesticides, what remained was a population consisting entirely of tough, resistant strains.

The number of generations produced by an insect population in a year is the primary factor affecting the speed at which they attain a level of resistance that renders the chemical obsolete. Flies in Canada, for example, have been slower to develop resistance to pesticides than those in the southern United States, where long, hot summers favor a rapid rate of reproduction.

While the resistance to chemicals that insects acquire illustrates similarities to our development of modern technologies, it also highlights an important fundamental difference between our cherished creations and evolution. An individual person can discover the next advancement in a technology, but evolutionary change isn't something that develops in an individual.

The example of insects reveals that the qualities an individual possesses at birth make them less susceptible than others to poisons and more likely to survive and produce offspring. Resistance, therefore, is something that develops in a population over time measured in several or many generations. New insect generations arise in a matter of days or weeks, so in their case, the changes happen relatively rapidly.

An important similarity between our complex innovations and evolution is greater change is possible in a shorter period with a large population rather than a small one. Had there been only ten insects in the population with half being resistant and half killed by the chemicals, they may have never acquired a resistance before natural factors eliminated their population. Five hundred thousand insects can experience adaptive change faster than 5,000. The more individuals, the bigger the gene pool and the greater the chance for an unusual advantageous mutation to happen. It's why animal breeders maintain large herds. They don't have to wait too long for desirable traits to occur.

Similarly, the more people on the planet, the more likely someone will make a discovery that results in the advancement of an existing technology.

The human population on the planet when electricity was first discovered was 100 million. At the time Westinghouse developed Tesla's motor, the population had increased to 1.5 billion people. Many of those people were academics or involved exclusively in research and development. These individuals had either partly or wholly devoted themselves to improving the technology. There is no counterpart in the evolutionary process to make organisms better or more advanced.

It is almost inconceivable that with so many people in the world, many devoted to nothing more than finding some way to improve existing technologies, that someone, somewhere doesn't come up with an occasional advancement. Our high-end technologies seem less a product of intelligence and more a product of the sheer immensity of the number of people in the world with time to devote to research and development.

Evolution doesn't seek to change or create complex organisms. It's nothing more than a biological mechanism that allows species to adapt to their environment. Evolution is a chaotic, random process with no intellectual capacity, yet it has created the nervous systems of mammals and the photosynthetic capabilities of plants, both of which make our telescopes, televisions, smartphones, and airplanes seem simplistic. If the complexity of organisms or technologies is the litmus test of intelligence, evolution, a completely random process, is "more intelligent" than humans.

We are an entity mimicking some aspects of evolution and rationalizing it as intelligence. All our advancements are the product of a burgeoning population, individuals devoted almost exclusively to the task, and the willingness to put 600 million years of the planet's natural resources on a credit card with a 300-year credit limit.

BRAINS

Our large brain has helped us to cure diseases, explore space, build large cities, overpopulate the planet, and use a disproportionate amount of natural resources. For someone living between 1850 and 2030, it would be easy to convince yourself of the advantages of having a large brain. The majority of humans began to enjoy the fruits of modern technology with the advent of medical care in the mid-19th century that resulted in a drastic drop in child mortality.

Many argue that the evolutionary development of a large brain is responsible for our superior intelligence and ability to commandeer and modify the planet. The oldest known fossil with a complex brain is about 520 million years old. The animal looked like a small crustacean with claws and seems to have had an elaborate brain-like structure consisting of a fore-, mid-, and hind-brain, all of which had specialized neural circuits. However, organisms with brains may have disappeared for periods of time.

Sea sponges did have neurons but experienced an evolved loss of these structures. Why would they lose their brain? One reason may be brains eat up an enormous amount of energy. In humans, up to 20 percent of our energy is spent feeding our brain. Sponges filter water and pick out only the useful nutritious particles. Adding a nervous system might not have helped with those tasks, and it became an unnecessary appendage and an evolutionary disadvantage. Sponges sitting on the seabed are good at what they do, and it doesn't require a brain.

In our natural tendency to give undue credence to complex theories, the most significant variable in our having a large brain may be one that is relatively simple and largely overlooked. At its most basic level, the ability to move differentiates organisms with brains from those without. There are only a handful of exceptions. All are organisms without brains living in aquatic environments that are able to move with the help of water. There are no known organisms with brains that don't move.

Generally speaking, the more complex an organism's movement patterns, the larger the relative size of the brain. Humans are the only organism that can run, walk, swim, dive, crawl, hang, swing, jump, throw, pound, lift, pull, grab, and climb. All of these movement patterns were made possible as the result of bipedalism and our unique shoulder anatomy. The evolution of our shoulders may have been the motivating force in our brain adapting to progressively larger sizes to accommodate increasingly complex movement patterns.

If brain size by itself resulted in superior intelligence, then Neanderthals would be the dominant species on Earth today. Their brains were slightly larger than ours. While they eventually died out, or more likely were the victims of genocide, it is entirely possible, even probable, Neanderthals could move more efficiently and possibly in more complex and diverse ways than their close cousins, *Homo sapiens.*

Researchers are currently doing backflips and somersaults trying to explain why, in the last 20,000 years, the average volume of the human male brain has decreased from 1,500 cubic centimeters to 1,350 cc, losing a chunk of our precious gray matter the size of a tennis ball. The female brain has shrunk by about the same proportion. No one knows for sure why the organ we are most prone to brag about has been shrinking.

Variables currently being explored as a means of explaining our diminishing noodle are an increasingly complex social environment, population density, climate change, and selection against aggression. Some scientists hypothesize that as complex societies emerged, people did not have to be as smart to stay alive. People have become increasingly specialized and reliant on others for most aspects of their survival. Some claim the wiring of our brain has become more efficient, while other authorities claim we are getting dumber.

While humans domesticated dogs, sheep, cows, and other animals, wheat and maize domesticated humans. Originally, they existed as wild grasses confined to small areas. Hunter-gatherers were living a fairly comfortable existence hunting and gathering until about 10,000 years ago.

A short time later, in nearly all parts of the world, humans were doing little from dawn to dusk other than taking care of grains and tubers. These plants had their human serfs clearing fields of rocks and pebbles so it might live a more comfortable existence. Human slaves weeded other plants that might encroach on their masters' space. They watched out for worms and blight, building fences and standing guard to provide protection. The servants of wheat and maize constructed ditches and canals to quench their thirst. Ruptured discs, arthritis, and hernias were a small price to pay for taking care of their sovereign. These plants manipulated humans to its advantage. By doing so, they made their own contribution to helping doom many complex life-forms on Earth.

In studies of some 30 animals that have been domesticated, similar to humans, every one of them has lost brain volume. This has typically been a 10–15 percent reduction relative to their wild progenitors. One thing is crystal clear: Domestic dogs don't move as efficiently as wolves. Domestic sheep don't move as precisely or powerfully as wild sheep. There is no domesticated animal that can or needs to move as well as their wild counterparts. Their survival is no longer premised on moving efficiently and in complex ways.

Today's domesticated humans are no different. We can't and don't move as well as our ancient relatives. A review of our natural history makes it clear that our movement patterns have simplified, become increasingly repetitive, and are greatly diminished.

The first and most significant change to our movement patterns was with the transition into an agricultural society. Planting, weeding, and harvesting fields require arduous, backbreaking work, but the movements are repeated over and over again, often for days on end. With the advent of the Industrial Revolution, factory workers were given simple, mind-numbing tasks that were often repeated over and over for a person's entire adult life.

In the last several millennia, we've gradually spent more and more of our waking hours sitting. The first chairs were documented about 5,000 years ago. Whether sitting in a vehicle, at work, to eat, or to be entertained, humanity is becoming increasingly proficient at dreaming up new ways of requiring sitting on our keister for extended periods of time. The average modern human is now sitting 6.5 hours a day. Recent science has provided extensive evidence that sitting is unhealthy. Ironically, to counteract the harmful effects of sitting, people now join gyms that provide a plethora of machines that allow them to sit while exercising.

Compare the movement patterns of domesticated humans to our hunter-gatherer ancestors. In most parts of the world, climbing trees was a daily activity for both men and women. It would have been necessary to harvest fruit or nuts, gain a height advantage to scan the skies for birds, spot prey or potential predators, or to escape danger lurking on the ground. Advanced age or injuries that prevented a person from climbing trees may have been a death sentence in areas with large, dangerous predators.

Digging up tubers, picking berries, hiking 20 miles to a new area where food might be more abundant, running from predators, making clothes and weapons, swimming to escape predators, fish, or collect other aquatic food, crawling for long distances to sneak up on potential prey, climbing up and down cliffs, and building a fire were but a few of the many varied activities. Many of the movement patterns of our ancestors integrated both the upper and lower body requiring greater mental acuity.

Today, climbing trees is an activity for children, assuming there are even trees where they live. Finding any person over the age of 30 who has climbed a tree without the aid of ladders, steps, or other modern conveniences would be akin to looking for a needle in a haystack. In the modern world, many consider anyone exercising more than 30 minutes a day as a fanatic. For those who do exercise, most isolate different muscle groups rather than moving in ways that integrate them. Vanity and aesthetics trump practical movement patterns in modern society and require less activation of our central nervous system.

Activities that require agility, balance, strength, endurance, hand-eye coordination and flexibility in some combination are almost exclusively the domain of children and young adults. As modern humans age, they quickly unlearn the movement patterns that have been evolutionarily programmed into our genetic makeup. Children are unlearning these movement patterns at earlier and earlier ages as concrete jungles replace real ones. The prevalence of neurotoxins and brain damage in infancy over the last five decades are further undermining our ability to reason.

Hunter-gatherers spent their time in more stimulating and varied ways and were less in danger of starvation and disease than early farmers. The average farmer worked harder than the average forager. The farm became the prototype of population explosions, new societies complete with pampered elite, inferior races fit for exploitation, wild beasts ripe for extermination, and a plethora of deities.

Despite the evidence, our fragile egos reject the notion of being outwitted by a grass. Admitting to being taken advantage of by a plant would shatter the foundation of the pyramid of myths humanity has carefully constructed over the last ten thousand years. It also flies in the face of our braggadocio regarding superior intelligence.

Living well during a 200-year period in the 4.5-billion-year history of the planet seems a flimsy rationale for claiming superior intelligence. Even if you expand the period of our supposed supreme intelligence through the Agricultural Revolution, 10,000 years is minuscule compared to the 3.7-billion-year existence of life-forms on Earth.

While our self-absorption allows us to bask in our success, our brevity on the planet will likely make our presence difficult to detect within a few million years. We will have maintained our place as a peak predator for somewhere between 10,000 and 20,000 years, depending on your point of view. Tyrannosaurs maintained their place at the top of the food chain for about 20 million years and saber-toothed cats for nearly 30 million years. If longevity is a primary consideration in evaluating intelligence, humans would be considered an organism well below average.

While vanity precludes an in-depth look in the mirror, a compelling case can be made that we are one of the dumbest species. How intelligent is an organism that destroys what it needs for life and then dies out? How intelligent is it to live entirely in an imagined reality, when the objective reality is what provides all the resources and sustenance it requires for survival? How intelligent is it to send humankind to the moon but fail to pay attention to Earth? Is being the only mammal ever domesticated by a plant really a sign of superior intelligence?

SUBJUGATION

We also rationalize our superior intelligence because of our ability and pre-disposition to kill, dominate, or use other organisms as we like. Using that theory, the most intelligent humans in history were Mao Zedong, Joseph Stalin, Attila the Hun, Genghis Khan, and Adolf Hitler. The serial killers currently locked up in maximum security prisons would be our current beacons of intellect. Just like Cortez was caretaker of the Aztec empire, humans often refer to themselves as caretakers of the planet.

We're one of the few animals that will kill another organism for no other reason than we can. There are people driving around every day in rural areas shooting badgers, porcupines, coyotes, wolves, birds, squirrels, and rabbits for no other reason than they can. People cut down trees and kill insects because they can. There are even people killing other people for the same reason.

The most likely scenario is we are neither more nor less intelligent than other species and rather unremarkable except in our own minds. Ultimately, we survive and reproduce to the best of our ability, the same as all other species. While our self-absorption and ego have us believing we are superior and more intelligent, it is no more than another creation of our active imaginations. There is no doubt that some humans are cleverer than others, but the same can be said for individuals of all species.

Hunter-gatherers picked and pursued wild plants and animals that were likely seen as equal in status to *Homo sapiens*. In fact, both status and intelligence were nonexistent issues. The fact that man hunted rabbits did not make rabbits inferior to man, just as the fact that lions hunted man did not make man inferior to lions. Organisms dictated their relationship with one another directly and negotiated the rules governing their shared habitat.

The first religious effect of the Agricultural Revolution was to turn plants and animals from equal members of an ecological round table into property, gods and man becoming the only important characters in the newly written script. Humans became the hub around whom the entire universe evolved. Gods explained why humans should dominate and exploit all other organisms.

Despite being tamed by a plant, humans and their gods became the sole beneficiaries at the expense of all plants and animals. All agricultural religions, including Christianity, Islam, Jainism, Buddhism, and Hinduism, found ways to justify human superiority and exploitation of animals. Animals stopped being viewed as living creatures that could feel pain and distress and instead came to be treated as machines. Everything we do to plants and animals is fueled by our perceived superiority and indifference.

The Agricultural Revolution ushered in the now universally held belief that humans are not a part of the planet's ecosystem. *Homo sapiens* is above it all—stirring, mixing, cutting, chopping, slicing, killing, and discarding whenever it suits our needs.

The new religious beliefs didn't stop with transforming other organisms from sentient beings deserving respect into mere property. It also began treating various classes of people as if they too were possessions. When ethnic groups or religious communities clashed, they frequently dehumanized each other. Depicting the "others" as subhuman beasts was a first step toward treating them as such. Those people who weren't at the forefront of the Agricultural Revolution, like Indigenous people, were soulless savages to those in the driver's seat.

Humans are born when many of their vital systems are still underdeveloped. Most bats and birds can fly when they are three weeks old. Human babies are helpless, dependent for many years on their elders for sustenance, protection, and education. Since humans are born underdeveloped, they can be educated and socialized to a far greater extent than any other animal.

Human babies emerge from the womb like a piece of dough. They can be molded, stretched, and shaped with a surprising degree of freedom. This is why parents, communities, and countries can train infants and children to become Mormon or Buddhist, fascist or socialist, warlike or peace-loving, Norwegian or Palestinian.

Contrary to popular humanist beliefs, science has found that human behavior is determined by hormones, genes, and synapses, rather than by free will—the same forces that determine the behavior of beaver, salmon, and northern spotted owls. Humans are subject to the same physical forces, chemical reactions, and natural-selection processes that govern all living beings. Humans are the outcome of blind evolutionary processes that operate without goal or purpose.

People believe our lives have greater intrinsic value than animals. People love telling themselves that we enjoy some magical quality that not only accounts for our immense power but also gives moral justification for our privileged status. The belief that humans have eternal souls and animals and plants don't explains why it is perfectly okay for humans to kill animals for food, for science, or even just for the fun of it.

However, there is zero scientific evidence that in contrast to trees, people have souls. Because there is no scientific middle ground, most people prefer to reject the theory of evolution rather than give up their souls. Many biologists and doctors go on believing in the soul but never write about it in scientific journals.

Who wouldn't want to believe in such an interesting and comforting story of humans acquiring an eternal spirit after a 3.7-billion-year-old recessive soul gene sprang to life at the beginning of the Agricultural Revolution? The lack of any supporting evidence has resulted in the life sciences ditching the soul.

Our judicial and political systems, in concert with various religious beliefs, largely sweep the absence of any tangible evidence of a soul under the carpet. We've constructed an imaginary wall between the department of biology and the departments of law, social, and political science which help us rationalize other organisms as inferior.

Humans are cognitive-dissonance gold-medal Olympians. We allow ourselves to believe one thing in the laboratory or when we visit the doctor and an altogether different thing in the courthouse, Congress, or church. We leave a revolving door in the wall between the disciplines that allows us to accept quantum mechanics, the theory of relativity, and all other scientific findings that don't put into question our soul. The evolutionary process is odd man out in this internal war between our imagination and what we view as an unpleasant alternative.

Humanity believes that if we can imagine something, we can both build and understand it. We are so delighted with our own ingenuity and intelligence; it gives us a false sense of knowledge and power. This power is problematic, as it takes us further and further from what is transpiring on the planet. Comfort and ease are the benefits we derive from being firmly ensconced in a world of our imagination.

What humans routinely ignore is how much we have in common with all organisms. To begin with, both literally and figuratively, all life on Earth shares a common ancestor. We still share 75 percent of our genetic material with mice, 73 percent with zebrafish, 50 percent with bananas, and 15 percent with mustard grass. Every organism shares the same evolutionarily acquired need to survive, reproduce, and grow.

Chlorophyll is present in all green plants and is responsible for the absorption of light to provide energy for photosynthesis. Hemoglobin is responsible for transporting oxygen in the blood of vertebrates. The similarity between chlorophyll in plants and hemoglobin in the blood of animals is striking. At the center of every hemoglobin molecule exists an atom of iron while at the hub of every chlorophyll molecule exists an atom of magnesium. In every other way, the molecules are identical. Chlorophyll is green blood designed to capture light. Blood is designed to capture oxygen.

There is a certain amount of freedom inherent in letting go of the belief we are the most intelligent organism. It allows a person to truly appreciate the unique skills and abilities of other creatures sharing the planet with us. We either forget or don't realize that the hearing, sight, and smell of wild animals, and especially of those animals that depend exclusively on these senses not only for food but also for self-preservation, are on a plane far and above that of civilized human beings.

It is unfathomable to us that organisms make decisions by processing information differently from humans and can be aware, sentient beings too. Our preconceived superiority ensures we routinely discount or completely ignore the, for lack of a better word, intelligence of plants and animals. How can they be intelligent when they are so different from us?

Rather than entertain the notion that other creatures might perceive and react to the world differently than us, humans have exactly two default modes. We either objectify them or attribute human traits, emotions, or intentions to them. In fact, we often utilize these same two mechanisms in our dealing with other people.

Women and minorities are often objectified to absolve the perpetrator of cruelty and abuse. The more we can demean and debase them, the easier it becomes to justify our brutality. Objectifying others, particularly during a time of war, is common. For example, the Assyrians conducted numerous military campaigns from the ninth century BCE to the seventh century BCE. They were known for their brutal tactics—depictions on their monuments showed conquered peoples being humiliated, tortured, or treated as subhuman. These behaviors have occurred throughout human history.

Many in the United States are horrified at our systemic racism of people with black and brown skin over the last 400 years, and rightly so. However, any *Homo sapiens* alive today, no matter the color of their skin, may have had ancestors who were racist. The ultimate form of racism is genocide. There were nine human species walking the Earth 300,000 years ago. Thirty thousand years ago, there were only three. By ten thousand years ago, all our cousins were gone. The timing of their extinctions strongly suggests *Homo sapiens* played a primary role in their demise.

Modern humans still have Neanderthal and Denisovan genetic material as part of our chromosomal makeup. The scientific euphemism is we "displaced" other species of hominids despite occasionally interbreeding with them. In much the same way the armies of Genghis Khan exterminated half of all Mongol tribes, it is likely our ancient ancestors "displaced" our cousins. Objectifying other organisms that are "different" from us, even if those dissimilarities are subtle, may be an evolutionarily acquired part of who we are as a species.

However cruel we are to people we view as different, humans up their game when it comes to other organisms. It's common in all cultures to hear people say, "They were treated like animals." Implied in the statement is no human should be treated with the cruelty that is both a normal and perfectly acceptable way of dealing with animals.

Most people are indoctrinated into objectifying animals and plants at an early age. They think of themselves as the subject and all other organisms as objects. Plants and animals are material things to be killed, eaten, mastered, destroyed, consumed, modified, or enslaved. The word *inhuman* is the ultimate non sequitur. Whether fueled by indifference or abject cruelty, we need only to look in a mirror to see *inhuman* in action.

The words *anthropomorphism* and *humanizing* describe attributing human traits, emotions, or intentions to other organisms. It helps us to simplify and make more sense of complicated entities. An entity is more likely to be anthropomorphized if it appears to have many traits similar to those of humans. Humanlike movements and physical features, such as a face, are two such triggers. Anthropomorphism renders an organism worthy of moral care and consideration. Anthropomorphized entities also become responsible for their own actions—that is, they become deserving of punishment and reward.

There are few people who fully objectify all other nonhuman organisms. Most people do some combination of objectifying and anthropomorphizing based on their own personal subjective criteria. There are literally dozens of animal rights groups, each with a slightly different idea of which animals deserve anthropomorphizing and which don't. I'm unaware of any group that protects the rights of spiders but plenty that seek to safeguard pets and livestock. Every person and every group draws their line in the sand of which organisms deserve their virtuous care and which don't. The lines in the sand are as numerous as people on the planet.

Scientists are just as culpable of this phenomenon as the rest of humanity. There is an infinitesimally fine line between being objective and objectifying. For those scientists studying sociology, psychology, and the physical sciences, it's easy to be objective. Everything becomes convoluted for scientists studying other organisms.

Students are taught that the power of science lies in its commitment to the ideal of objectivity. The goal is to design, conduct, analyze, and interpret experiments with dispassionate, robotic, and uncompromising devotion. The heartless, soulless interrogation of nature becomes scientific nirvana.

Scientists who don't objectify the organisms they study open themselves up to criticism from both the public and their colleagues. Imagine becoming a wildlife or botanical researcher because of a love for animals and plants, but then the profession forces you to objectify the organisms that drew you to the vocation in the first place. Most researchers conform to the pressures of colleagues in their chosen line of work to ensure their research is taken seriously and they become successful in their chosen field.

In 2015, Christopher Filardi of the American Museum of Natural History captured the first known male moustached kingfisher in Papua New Guinea and promptly killed it for a museum specimen. While he acknowledged that the future for the species is far from secure, the rationale he and his colleagues gave was they assessed the state of the population and concluded it was healthy enough to take the specimen. His actions were completely consistent with the standard practice of field biologists.

Many authors of scientific papers who defended Filardi's actions used the descriptor *responsible* scientific collecting, but no one defines *responsible*. It's déjà vu all over again. Like people's consistent self-assessment of being good drivers or intelligent beings, it's easy to be a responsible scientist when you're the one who defines it.

Filardi collected the bird with the best of scientific intentions. There are plenty of scientists collecting individuals of rare species, often knowing or caring much less about the status of a population than he did. Every one of them does so sincerely believing they are responsible scientists.

The majority of the scientific community's worldview of human superiority is no different than people of all disciplines since the dawn of the Agricultural Revolution. The collected bird was nothing more than an object of study that promised information for future scientific papers. Filardi is the subject, the moustached kingfisher the object. Nonhuman organisms scrum around in the messy ecosystem called Earth. Humans no longer view ourselves as a part of nature. We're above it all, modifying and destroying when it meets our needs, whims, and desires.

Similar to the rest of humanity, I was also raised to objectify animals, plants, and even other people. My adult life has been spent trying to jettison those teachings without flipping over to humanity's other default mode of humanizing them. I've achieved a modicum of success but will be pushing up daisies long before completely overriding my preadult programming. Striving to emulate the way ancient hominids viewed other organisms as neither inferior nor superior seems a worthy goal. For most people, the scrutiny required to critically examine early teachings is too painful, time-consuming, and uncomfortable to even contemplate. It's easier and more pleasant to accept on faith information taught to us when we are young.

In three decades of research and monitoring wildlife populations, it didn't even cross my mind to collect a specimen. I tried to do everything possible to minimize my impact on both individuals and populations. Adding one human-induced impact to the myriad they regularly faced seemed the definition of uncaring arrogance. In today's world of biological science, conceit almost always wins over humility.

Philanthropists are seen as good people because they give money to those who are less fortunate. In much of the world, there is a season of giving every year where individuals, communities, and organizations come together with a commitment to help those in need. Some religions regularly give to the poor and disadvantaged. Humans generally see giving as a good thing and taking from those less fortunate as bad. When it comes to our planet, we do the exact opposite.

When plants and animals parted company on their evolutionary journey millions of years ago, plants became the givers and animals the takers. But plants call the shots, make the air, and eat sunlight. Without plants, there would be no animals and certainly no humans. What is increasingly lost on today's humanity is that without them, there is nothing.

Instead, humans have waged war on the planet's grasslands and forests. In just a few centuries, humanity has destroyed all the major forests, wetlands, and grasslands around the world, leaving but tiny remnants. People are the quintessential takers. Biting the hand that feeds us isn't enough; we murder the givers.

PLANTS AND ANIMALS

Feral pigeons are also called city doves, city pigeons, or street pigeons. They are derived from domestic pigeons that were originally bred from the wild rock dove. They are the avian equivalent of the feral horses roaming western rangelands. Rock doves naturally inhabit sea cliffs and mountains. Feral pigeons find the ledges of buildings as an adequate surrogate for sea cliffs and have readily adapted to urban life. These birds are largely considered a nuisance.

I occasionally find myself in downtown Portland, Oregon. For many years there was a food court that took up most of an entire city block. At lunchtime, there were lines on the sidewalk at some of the more popular vendors, resulting in various choke points to pedestrian traffic moving along sidewalks. It's a scenario that plays itself out repeatedly in many cities around the world. It is crowded, chaotic, and requires a little patience and occasional bumping and jostling to successfully navigate through and around the lines.

I had stopped at one of the vendors for a grilled cheese sandwich and after ordering stepped into a small nook in between two food vendors to await the arrival of my lunch. It afforded me time to not only watch but actually see. A number of pigeons were taking advantage of the bounty provided by messy eaters.

There was a half dozen pigeons walking on the sidewalk amid this mass of humanity within my line of sight. They were doing a much better job of navigating the human traffic than the people were. In their case, a single error in anticipating what the pedestrians were going to do would have resulted in serious injury or death. They not only were avoiding being stepped on but were successfully scavenging for morsels at the same time. For all practical purposes, the pigeons were completely invisible to the people traversing the sidewalks.

Taking a peek at what pigeons do through the human lens using our own definitions of intelligence is fascinating. The first definition is accomplishing complex cognitive feats. Avoiding being stepped on, anticipating the actions of multiple people simultaneously, and scavenging for food seems to meet that definition in spades.

The ability to achieve goals: The pigeons were successfully getting enough to eat from people due to the people doing a poor job of juggling both a lunch and a smart phone or trying to eat and walk at the same time.

The ability to resolve genuine problems or difficulties: I found the crowds uncomfortable and either avoided the area or tried to stay out of pedestrian traffic to the extent I could. The pigeons, however, seemed totally at ease despite being a single human step away from disaster.

Acquiring, storing, retrieving, combining, comparing, and using information and conceptual skills in new contexts: The pigeons probably acquired the skills of operating in heavy human traffic by at first working around the perimeter or being present when the sidewalks weren't teeming with activity. They gradually learned how to anticipate human movement patterns and deal with an increasingly bustling sidewalk. Maybe the most amazing element of their behavior was anticipating multiple people's actions simultaneously better than other humans.

Think rationally: check. Deal with the environment: check. Adapting to circumstances: It's a long way from sea cliffs and mountains to the sidewalks and food courts of our increasingly urban environment. Cities sprang up several hundred years ago, sidewalks 100 years ago, and food courts probably about 50 years ago. Pigeons have successfully adapted to the environmental changes precipitated by human development. Under all of the definitions of *intelligence* created by humans, these pigeons were operating in a much more dangerous environment at a level often exceeding that of humans.

In addition, pigeons can find their way home by sensing Earth's magnetic fields. They can also use cues based on the position of the sun.

Here are these marvelous birds we consider pests operating successfully in the midst of an artificially created environment that are for all practical purposes invisible to us. They illustrate that we can't see anything that doesn't look like us. We wouldn't be able to recognize intelligence in another organism if it slapped us in the face or, in this case, quite literally walked around right under our feet. Everything alive that isn't human is just beyond our field of view.

Imagine if the roles were reversed and pigeons had their own subjective definition of intelligence. It might look something like this: Organisms that can be blindfolded and released 1,000 miles from home and find their way back would be considered highly intelligent. Those who were released 500 miles from home with no blindfold and could make their way home were of average intelligence. Those who could only find their way back home from a 50-mile destination were morons. Imagine what they would think of humans, many of whom may even have trouble finding our own vehicle where we left it mere hours before in the shopping mall parking lot.

Pigeons would probably be impressed with sockeye salmon who, after hatching, travel from a lake via connecting streams and river to the ocean, which can be a distance of hundreds of miles. Once they reach the ocean, they might travel an additional 1,000 miles to reach their feeding grounds.

Like pigeons, the salmon use Earth's magnetic field like a compass, along with chemical cues and possibly other senses we don't yet understand, to return to their natal streams and lakes after spending several years in the ocean.

They would also be impressed with many types of bats who utilize an active system of orientation called echolocation. These animals use vibrations in the air to detect obstacles and other objects, using information from both the original signals or calls they emit and echoes of these sounds returning from nearby surroundings and potential prey. The differences between the sound they produce and the echo they hear is the data they use to echolocate. They vary their call duration, time between calls, and the frequencies of their signals to maximize the information they obtain from returning echoes. Sound travels more than 1,100 feet per second in air, so bat echolocation involves split-second timing.

Bats cannot broadcast and receive signals at the same time because outgoing calls deafen them to the faint returning echoes, so they separate pulse and echo in time. As an echolocating bat closes in on its insect prey or a wall in a dark cave, the pulse of its calls must be shorter and shorter to minimize overlap between outgoing calls and returning echoes. The bat is processing and reacting to information at a rate that would make the reflexes of a National Hockey League goalie seem like that of a vegetable.

While this is usually catalogued by humans as interesting, it certainly doesn't reach our threshold for considering them intelligent. How could it? They don't look like us, and we can and do kill them without much pause.

While it is unlikely humanity will be around long enough to know for sure, the most intelligent organisms on the planet may be forests. Within those organisms, the most intelligent individuals are the ancient trees. Researchers have already discovered the largest trees in forests act as central hubs for vast below-ground mycorrhizal networks. These trees support seedlings by infecting them with fungi and supplying them with the nutrients they need to grow. They even change their root structure to make room for the baby trees. They also make decisions as to which of their seedlings should be helped and which shouldn't. Mats of mycorrhizal fungi link trees into gigantic smart communities spread across hundreds of acres. Together, they form vast trading networks of goods, services, and information.

Research also recently discovered deciduous and coniferous trees trading nutrients over the course of a season. This cooperative underground economy appears to lead to healthier forests, greater resilience in the face of disturbance, and more total photosynthesis. Forests have had their own complex underground internet for millions of years before humans even appeared on the planet.

In retrospect, why should it come as a surprise that organisms that have been on the planet for hundreds of millions of years longer than humans have developed an awareness, patience, and decision-making abilities far exceeding our own? While humans unsuccessfully aspire to simultaneously live independently and cooperatively with other people and organisms, forests and trees have been doing it for millions of years.

We've already annihilated most of the forests and are doing everything possible to decimate those that remain, dooming ourselves in the process. We will cease to exist long before we know even a tiny fraction of what forests have already figured out.

Our narcissism prevents us from even considering another organism might know more than us. In fact, this belief transcends other organisms and includes an earlier version of ourselves.

A popular insurance television commercial made a sales pitch that using their website is "so easy even a caveman could do it." It's an amusing and effective sales campaign because people need no convincing they are much more intelligent than our ancestor hunter-gatherers.

For all those people who believe they are superior to pre-agricultural humans, try living without modern clothing, equipment, or weaponry in the backcountry for a week—anywhere, anytime. It will give you an appreciation for how incredibly tough, aware, talented, clever, agile, and dexterous they were.

People's undying faith in their intelligence and superiority has many believing that somebody, somewhere will magically come up with an antidote to overpopulation and overexploitation of our environment. Jettisoning false expectations about humanity's capabilities and intelligence can result in not being as disappointed by what is now occurring daily, both on large and small scales. In fact, I now fully expect actions that will make humanity increasingly vulnerable to the limitations of the Earth's environment and am rarely disappointed.

The brief golden age of the last several centuries is the culmination of several million years of rapid behavioral changes in *Homo sapiens*. Our transformation has sown the seeds of catastrophe. The consequences will be dire. Our accomplishments have come at the cost of our planet's biodiversity. The material wealth that shields us from disease and famine was acquired on the backs of forests, dry land ecosystems, marshes, and clean water and air, not to mention the unprecedented cruelty to animals. Despite that, there is no evidence that all the environmental destruction over the last 12,000 years has increased human happiness. In fact, likely the opposite is true.

Everybody is a genius, but if you judge a fish by its ability to climb a tree, it will live its whole life believing that it is stupid.

—Albert Einstein

CHAPTER 3

CANARIES

Canaries were once used as an early-warning system in coal mines. Toxic gases like carbon monoxide or methane would kill the canary before affecting the miners. Signs of suffering in the bird would alert miners to hazardous air quality giving them time to escape mine shafts and adits.

What the canary was to coal miners is what fish, wildlife, and native plants are to the health of the planet. Other organisms provide the biodiversity necessary for both our and their survival. These creatures are often more sensitive to the impacts humanity makes to the climate, water, air, and land than we are. They are a ubiquitous canary to our actions. Their struggles are a clear forewarning of a coming tragedy for the planet, other organisms, and humanity. The repeated notices ignored by humanity over the last two centuries are like lighthouse beacons transformed into tiny candles.

EASTERN NORTH AMERICA

Biologists describe plants or animals on which other organisms in an ecosystem depend as keystone species. They hold the ecosystem together and play a disproportionate role in determining not only which plants and animals can survive in an area, but their relative abundance. They provide the foundation on which the ecosystem is built. If those species are removed, the ecosystem changes drastically.

The passenger pigeon was a keystone species in much of eastern North America for many thousands of years. They may have been the most abundant bird in the world. Passenger pigeons had a small head and neck with a long, wedge-shaped tail. Their long, pointed wings were powered by large breast muscles that gave the capability for prolonged flight. The bird was physically adapted for speed, endurance, grace, and maneuverability. Passenger pigeons were an extremely fast flying bird reaching speeds of 62 miles per hour.

The male's head and neck were clear bluish-gray with black streaks on the shoulders and wings. There were patches of pinkish iridescence at the sides of the throat merging into shining metallic bronze, green, and purple at the back of the neck. The lower throat and breast were a soft rose, gradually shading to white on the lower abdomen. The irises were bright red. The bill was small, black, and slender, and the feet and legs were red. The female's head and back were a brownish gray. The iridescent patches on the throat and back of the neck were less bright than the males. The breast was a pale cinnamon-rose hue.

At the time of the European discovery of America, there were between 3 and 5 billion passenger pigeons. During migration, flocks up to a mile wide and hundreds of miles long could take days to pass overhead. Genetic studies have demonstrated that prior to the bird's 19th-century decline and eventual extinction, the population size of the passenger pigeon was stably abundant for tens of thousands of years. It was even abundant during times when deciduous trees were rare and ice sheets covered half of its known range. Insights gained from modern genetic analysis show passenger pigeons thrived and adapted to significant environmental changes. To put the number of passenger pigeons into perspective, there were as many passenger pigeons in eastern North America as there are email users in the world.

There were so many passenger pigeons that they functioned like a tornado and wildfire in their effect on the forests they called home. The incredible number of passenger pigeons and overcrowding resulted in them breaking all the branches off small clusters of trees, killing them. The huge deposits of guano temporarily snuffed out vegetation.

The storms of passenger pigeons created small openings in the forest, allowing sunlight to reach the forest floor. Where that occurred, it powered the growth of thick patches of understory vegetation. Sunlight stimulated a variety of flowering plants that attracted pollinators and animals that feed on plants. Reptiles and amphibians also reaped the benefits of the warm sunlight and abundant insects. The thick ground vegetation in the small openings bestowed cover for small mammals to raise offspring, ultimately providing prey for hawks and owls. The edges of these small openings made it possible for nesting songbirds to locate and avoid predators.

As the small formerly forested areas that had been killed by the passenger pigeon tempest regenerated and the upper layer of mature trees reestablished, the animal community changed, creating homes for bats, owls, squirrels, and woodpeckers. These forest inhabitants found shelter and security in the canopy of large trees but still had the means and opportunity to hunt and forage in the nearby open areas. The diverse forest vegetation created by the roosting, feeding, and nesting passenger pigeons provided living accommodations for an incredible array of plants and animals.

Patches of mature oaks, chestnuts, beeches, and American elm provided ample seed crops for sundry birds and animals and attracted passenger pigeons to return and restart the regenerative cycle once again. Dense concentrations of nesting passenger pigeons generated ecological hot spots of biodiversity and bioabundance.

Pigeon nesting areas were an incredible short-term, protein-rich buffet for hawks, owls, weasels, raccoons, foxes, skunks, and other predators. The passenger pigeon's technique of survival was based on mass tactics of overwhelming local predators. Flocks often numbered hundreds of thousands of birds. When a flock this size established itself in an area, the number of local animal predators was so small compared to the total number of passenger pigeons that little damage could be inflicted on the flock as a whole.

Since accurate data of such an immense number of birds was neither practicable or a priority at the time, it is only possible to give estimates on the size and population of nesting areas. One large nesting in Wisconsin was reported as covering 850 square miles with an estimated 136 million passenger pigeons. The birds were so jam-packed in these areas that many nests could be counted in a single tree.

A single, white, elongated egg was laid per pair. The duties of incubating the egg and feeding the young were a joint endeavor by both parents. The young, called squabs, grew and developed rapidly. After 14 days, they had grown to the size of adults. The squabs soon fluttered to the ground to hunt for their own food. Once the squabs' wings fully developed so flight was possible, they were assimilated into the flocks. Before the cold weather forced their seasonal migration south, passenger pigeons moved about randomly in northern forests in late summer.

Passenger pigeons were occasionally pursued for meat by Indigenous people, but hunting intensified after the arrival of Europeans, particularly in the 19th century. Commercial hunters began netting and shooting the birds to sell in city markets, as live targets for trap shooting, and even as agricultural fertilizer. Once pigeon meat became popular, hunting began on a colossal scale that lasted for many decades. The flocks were so dense that killing massive numbers of the birds was possible with little difficulty.

The bird painter John James Audubon described the slaughter at a known passenger pigeon-roosting site:

> Few pigeons were then to be seen, but a great number of persons, with horses and wagons, guns and ammunition, had already established encampments on the borders. Two farmers from the vicinity of Russellville, distant more than a hundred miles, had driven upward of three hundred hogs to be fattened on the pigeons which were to be slaughtered. Here and there, the people employed in the plucking and salting what had already been procured, were seen sitting in the midst of large piles of these birds. The dung lay several inches deep, covering the whole extent of the roosting place.

The survival strategy of overwhelming predators with sheer numbers worked for millennia but became perilous when men began hunting them. Passenger pigeons were massed together, especially at nesting sites. Hunters slaughtered them in such huge numbers that they were shipped by the boxcar-load to the eastern cities. In New York City, in 1805, a pair of pigeons sold for two cents. Slaves and servants in 18th- and 19th-century America often saw no other meat.

Similar to some species of bats, passenger pigeons were highly social. They practiced communal breeding with up to a hundred nests in a single tree. Once their numbers were reduced below a certain threshold, the deteriorating social structure likely would have also contributed to their eventual extinction. Passenger pigeons could not adapt themselves to existing in small flocks.

Between 1800 and 1870, passenger pigeons suffered a 97 percent population loss. Even then, there were still an estimated 136 million breeding adults in North America in 1871. An 1855 account from Ohio described a "growing cloud" of passenger pigeons that blotted out the sun.

After the Civil War, the expansion of the telegraph and the railroad enabled a commercial pigeon industry to flourish. Professional hunters learned quickly about new nestings and were able to follow the flocks by train. One of the last large nestings of passenger pigeons occurred in 1878 near Petoskey, Michigan, where 50,000 birds were killed every day for nearly five months.

The pigeons were shot, trapped in nets, asphyxiated with burning sulfur, poisoned, and their roosts were torched. A Potawatomi leader described witnessing a massacre in 1880, watching scorched adults fleeing and young, unfledged birds bursting open upon hitting the ground. I cannot help but think, as the Potawatomi leader surely did too, of the Potawatomi's own experience with death at the hands of early American settlers as a result of a forced removal from Indiana by militia in 1838. The forced march of the Potawatomi from Indiana to Kansas later became known as the Trail of Death.

The intense hunting of passenger pigeons resulted in a final sharp decline between 1870 and 1890. Despite the commercial slaughter, the birds might have survived into the early 20th century if hunters weren't simultaneously disrupting their nesting grounds. Even as the pigeon's numbers crashed, there was no effort to save them. People hunted and slaughtered them with even greater effort until the very end. On March 24, 1900, a boy in Pike County, Ohio, shot the last recorded wild passenger pigeon. Fourteen years later, on September 1, 1914, the last known passenger pigeon died at the Cincinnati Zoo.

The passenger pigeon wasn't the only major casualty in eastern North America during this period. The American chestnut tree once dominated the eastern half of the United States and Ontario, Canada, numbering between 3 and 4 billion. They were a rapidly growing deciduous tree reaching 90 to 100 feet tall and a diameter of over nine feet. The tree's huge population was partially due to a combination of rapid growth and a large annual seed crop. Nut production began when the trees were seven to eight years old and was reliable in comparison to oak trees, whose acorn numbers fluctuate wildly from year to year.

The American chestnut was a prolific bearer of nuts, usually with three nuts enclosed in each spiny, green burr lined in tan velvet. The nuts developed through late summer with the burrs opening and falling to the ground near the first fall frost. The American chestnut was important for wildlife, providing much of the fall mast for white-tailed deer, wild turkey, black bears, and passenger pigeons.

The nuts were high in fiber, vitamin C, protein, and carbohydrates and low in fat. They returned more nutrients to the soil in the fall than other trees because of elevated levels of nitrogen, phosphorus, potassium, and magnesium in their leaves. This helped the growth of other plants and microorganisms to reboot forest vitality.

The chestnut blight slipped into North America from Asia in the wood of Chinese chestnuts destined for fancy gardens. In 1904, a tree in the

Bronx Zoological Park became infected. Leaves curled and scorched the color of cinnamon, and rings of orange spots spread across the swollen bark.

Within a year, orange spots flecked chestnuts throughout the Bronx, fruiting bodies of a parasite that had already killed its host. Every infection released a horde of spores with the rain and wind. A researcher at the New York Botanical Garden identified the killer as a fungus new to man.

The chestnut blight jumped dozens of miles a year across Connecticut and Massachusetts. These giants of the eastern forests succumbed by the hundreds of thousands. The country watched stupefied as New England's priceless chestnuts melted away. Called the sequoia of the East, every fourth tree of a forest stretching 200 million acres from Ontario down to the gulf was doomed.

Like an earthquake in the ocean, the demise of the chestnut sent a tsunami of effects through the forest community. It resulted in the extinction of seven moth species and a precipitous decline in squirrel populations. The lack of squirrels had a cascade effect resulting in fewer hawks, red wolves, bobcats, fox, bears, and cougars. It decreased the abundance of cavity-nesting birds and even negatively affected insect populations in streams and lakes and the fish that utilized them. The profusion of chestnut bloom supported native insect pollinators. One of the greatest ecological disasters to strike the world's forests in all of history had come to fruition.

The once-majestic American chestnut now survives as little more than a bush by sending up stump sprouts that inevitably succumb to the blight and die back to the ground. In cruel irony, the blight doesn't kill the root system of trees, so they still exist in the form of mere wisps of their former selves. Because they are no longer capable of bearing seeds, many in the scientific community describe them as functionally extinct. Today, few people remember that chestnut trees dominated the eastern forests from Ontario, Canada, to Indiana, to Mississippi, to Georgia.

Bottomland forests in eastern North America were once populated by the American elm, a common tree providing much of the upper forest canopy. The geographic range of the American elm and passenger pigeon nearly mirrored each other. Chestnut and American elm trees provided many of the roost and nest trees for the vast numbers of passenger pigeons. They also provided a reliable food source at critical times during migration, nesting, and overwintering.

American elm reached maturity in about 150 years, and many trees lived to be 300 years old. The trunks of mature trees were between three and six feet in diameter. With their arching branches and stately shape, American elms provided tunnels of cooler shade in town squares and city streets across the United States. It was tolerant of urban areas, grew rapidly, provided abundant shade, and had a graceful form. The American elm was the most popular tree planted in the booming cities of the 19th century. By the 20th century, many streets were lined with only elms and were shaded in summer by a cathedral-like ceiling of their branches spreading like fountains. Their canopy was almost as wide as they were tall.

With eerie similarities to the chestnut blight disaster, in 1931 a furniture company in Cleveland, Ohio, unwittingly bought logs infected with Dutch elm disease from France. Invading the water-conducting vessels of the tree, the disease shows its earliest external symptoms of infection as yellowing and wilting leaves on individual branches. These leaves often turn brown and curl up as the branches die. Although initially only a part of the tree canopy may be affected, the symptoms progress rapidly throughout the entire tree. Highly susceptible trees often succumb in a single year, but others may linger for several years before dying.

The disease quickly spread east from Ohio and within two years was affecting American elm trees in New Jersey. It was found around the port of New York City in 1933. Quarantine and sanitation procedures arrested most cases within a 150-mile radius of metropolitan New York City until 1941 when the demands of World War II curtailed efforts to prevent spread of the disease.

Dutch elm disease stormed west and south from New England, reaching the Detroit area in 1950. It progressed to Chicago by 1960 and

Minneapolis by 1970. Neither American nor European trees had any resistance to the disease caused by yet another fungal terror introduced from another continent. The larger the tree, the more susceptible they were to the disease. Of the estimated 77 million elms in North America in 1930, over 75 percent had been lost by 1989.

In an almost identical scenario to chestnut blight, American elm trees can survive the fungus when they are young and small but become susceptible to the disease as they mature. The beautiful large trees are gone forever from the native bottomland forests originally found throughout Eastern and Central North America.

Like hundreds of species who have either mostly or entirely disappeared from the Earth, we'll never fully understand the role elm played in the native woodland community and the food web. They were the first trees to flower in the spring, often when there was still snow on the ground. Their seeds mature in late spring, a time when there are few other seeds available to many birds and mammals. A number of native insects also relied on elms for food and cover. It appears passenger pigeon migrations northward coincided with American elm seed formation.

The American elm remains the state tree of Massachusetts, Nebraska, and North Dakota. A grizzly bear known as Monarch, killed in 1922, was the last confirmed sighting of this species in California. The grizzly bear was designated as the official state animal in California in 1953, the symbolism is a reminder of what was and is no more.

Hunting was the primary driver of the passenger pigeon's passage into extinction, but widespread deforestation also played a significant role by destroying its habitat. The birds needed huge forests to thrive. Forest clearing and farming were diametrically opposed to their survival. Early settlers cleared the eastern forests for farmland. The birds were forced to shift their nesting and roosting sites to the forests that remained. As their forest food supply declined, the birds began utilizing the grain fields of farmers. The farmers retaliated by shooting the birds and using them as a source of meat.

Forest cover in the eastern United States reached its lowest point in roughly 1872 with about half the amount as existed in 1620. The rapid clearing of natural forests for agriculture and other uses would have led to the same inescapable outcome for the passenger pigeon.

Except for logging, the introduction of first the chestnut blight and then Dutch elm disease are the most significant events in the history of native forests in North America. The elm seeds passenger pigeons relied on early in their northerly migrations would have disappeared as well as the chestnut mast in late fall. The birds would have had to shift their nesting and roosting sites from chestnuts and elms to trees in the forest only marginally suited to the task, potentially damaging those that remained. With little food available for nesting and wintering, reproductive success would have plummeted along with their population.

It was a virtual horse race, with hunting barely beating forest conversion, and chestnut blight and Dutch elm disease by a nose, in driving the passenger pigeon to extinction. Alternate-reality novels like Stephen King's *1962* or S. M. Stirling's *The Change* series can be both fun and thought-provoking, so let's imagine this story. In an alternate reality of our own making where not a single passenger pigeon was shot, forest conversion and logging would have given the species the ax, pardon the pun, by the 1930s. Had European Americans never done any logging, hunting, or land clearing but only introduced the chestnut blight and American elm disease and then left the continent to its own devices, it is still likely the passenger pigeon would have been driven to extinction. The rapid loss of the American elm and American chestnut, two species it had evolved with and relied on for food, nesting, and roosting, would have probably been too much to overcome.

In the extinctions of the Carolina parakeet, ivory-billed woodpecker, and Bachman's warbler in roughly the same geographic area, the horse race ended differently. Forest conversion beat hunting by an equally thin margin of victory. The loss of forests was the most significant factor chaperoning these birds to their extinction.

WESTERN NORTH AMERICA

While the passenger pigeon filled the role of keystone species in eastern North America, the beaver did the same in the West. In fact, beaver helped shape the landscape. They are also referred to as ecosystem engineers because they physically alter habitats by cutting down trees, constructing dams, and building lodges. In doing so, beaver indirectly changed the distribution and abundance of many other animals and plants.

Beaver are the largest rodent in North America, reaching up to 90 pounds. They are highly adaptable and can be found from boreal areas in northern Canada and Alaska to arid lands in the western United States and northern Mexico. Water and wood are the only materials necessary for their creative ecosystem architecture and construction activities.

Prior to the arrival of Europeans, there was an estimated population of between 60 and 400 million beaver on the North American continent. There were at least 25 million beaver dams in small- and medium-sized streams. About 75 percent of beaver live in lodges; the remainder burrow into banks.

Beaver have stocky bodies with a yellow-brown to almost black coat and a broad, flat, scaly tail. Their large, orange incisors grow continuously throughout their ten-to-twelve-year lifetime. Beaver regularly move between water and the adjacent land. A beaver's movements on land are awkward, making them vulnerable to predators. In water, however, beaver can swim up to six miles per hour and stay submerged for up to 15 minutes while traveling over half a mile. The beaver's tail is used as a rudder in swimming, as a balance prop while working on land, and as a signal of danger when slapped on the water. They have an acute sense of hearing, with valves in their ears that close while submerged.

Beaver eat leaves, woody stems, and aquatic plants. Their chief building materials are also their preferred foods and include aspen and willow. In areas with cold climates where they live in North America, they spend the winter inside their lodge chamber, feeding on branches they have stored on the muddy pond floor as a winter food supply. The water acts as a refrigerator, keeping the stems cold and preserving the nutritional value.

Beaver form strong family bonds and are social animals. It's generally believed beaver pair for life. Each group is made up of one breeding pair, the year's kits, and the surviving offspring from the previous year, called yearlings. These family groups live together in the winter and share food from their communal stored supply.

The largest beaver dam ever recorded was found in 2007 in Alberta, Canada, and was longer than Hoover Dam, a little over a half mile. Historically, the incredible number of beaver dams accomplished many of the same functions that irrigation and flood control dams do today. They slowed and delayed water from floods caused by storms and melting snow. After decimating beaver populations, humans have spent billions of dollars on the construction and maintenance of dams trying to mimic what beaver did naturally.

The pools beaver created saturated the surrounding areas, making them suitable for water-loving plants. The water in the beaver ponds and the lush vegetation attracted a greater variety and numbers of wildlife. During summer months, the water in the saturated areas would slowly seep back into the channel providing permanently flowing water when streams would otherwise be low or even completely dry. Beaver, in essence, allowed fish and their young to survive in many streams when they would have otherwise perished. Beaver ponds provided winter habitat for trout and salmon when temperatures were frigid, being the only place in the streams that didn't freeze to the very bottom.

Beaver ponds created habitat for waterfowl and amphibians. They also supported insect species normally only found in lakes. Because a drainage with beaver dams would have free-flowing streams stair-stepped with beaver ponds, they provided a greater diversity of insects, birds, mammals, amphibians, plants, and trees.

The fur trade began in the 1500s as an exchange between Indians and Europeans. The earliest fur traders in North America were French explorers and fishermen who arrived in what is now eastern Canada. The Indigenous hunters traded furs for such goods as tools and weapons.

Free-roaming trappers have been romanticized in tales of mountain men. The reality was beaver trapping was arduous, usually frigid, and sometimes dangerous work. Most trapping was conducted during winter months when the animal's fur grew thickest. Beaver traps were usually set in ponds near beaver dams. Trappers often waded through thigh-deep mud overlaid with freezing water to set their traps. Pulling a foot and leg out of the mud in a beaver pond to take a single step could make the most taxing exercise in modern big-box gyms seem like a stroll in the park.

Trappers set traps underwater, anchoring it with a stake tethered to a short chain. The stake was long enough to be seen above the water line. To attract beaver, trappers used castoreum oil, taken from the musk glands of beaver. The yellowish substance is used by beaver to scent mark their territory or attract a potential mate. Trappers rubbed it on top of the anchoring stake or suspended it from a tree branch in a small bag above the trap.

At night, an unsuspecting beaver would swim toward the odor, and inspecting the spot with the scent, step on the trap's trigger. Once the steel-jawed trap catch was released, the jaws would slam shut around the beaver's leg. Frightened and in pain, the beaver would swim toward the deepest water available. It was a flight response that served them well over numerous millennia to help them escape dangers originating from land. However, once in deep water with an anchor firmly attached to its foot, the trapped beaver would eventually tire and drown.

Trappers preferred drowning beaver so that their catch would be preserved from scavengers as long as needed until they checked their trap. Trappers would remove the dead beaver and reset the trap. This continued until they had vacuumed up all the beaver in an area. During the early 19th century, they would move on only after ensuring their competitors were left with nothing.

The fur trade gained the interest of Prince Rupert of England in the 17th century, who in 1668 sent ships to the New World. On May 2, 1670, the Royal Charter granted exclusive trading rights of the Hudson Bay watershed to "the Governor and Company of Adventurers of England trading into Hudson Bay." The company used James Bay and Hudson Bay as their hub of operations for nearly a century. Annually, trappers brought furs to barter for manufactured goods such as knives, kettles, beads, needles, and blankets. By the late 18th century, competition forced Hudson Bay Company to expand its operations into the interior of the continent. Several of the posts the company established eventually morphed into what is now Winnipeg, Calgary, and Edmonton.

By the last quarter of the 18th century, as the plains nomads accumulated beaver pelts for trade with the Europeans, they steadily exterminated the beaver from the upper Missouri River basin. In 1804, Lewis and Clark remarked that on the Niobrara River, "Beaver-houses have been observed in great numbers on the river, but none of the animals themselves." The Lewis and Clark expedition to the Pacific Ocean in 1805 and 1806 was the precursor to the development of fur trading in the western continental United States. Several companies competed heavily for this western trade, including John Jacob Astor, who founded what is now Astoria, Oregon, at the mouth of the Columbia River.

The competition for beaver between rival trading companies in the Pacific Northwest and the Northern Rocky Mountains in the continental United States was fierce. In 1820, for example, the Hudson Bay Company, outraged over an 1818 treaty that established joint occupancy of the Columbia Basin with the United States, ordered its trappers to do their best to eliminate beaver from all of the territory south and east of the Columbia River in the Snake River region. The company expected the Americans to try to occupy the area and believed eliminating beaver would discourage American fur traders from occupying an area the company considered its own.

Between 1819 and 1823, the British had 60 men trapping the Snake River country, taking out 80,000 beaver with pelts weighing a total of 80 tons. Nearly 15 years later in 1835, Nathaniel Wyeth led an expedition that founded Fort Hall near present-day Idaho Falls, Idaho. His journal chronicled large areas of the Snake River plains downstream of Fort Hall in the mid-1830s as having no beaver because the Hudson Bay Fur Company had trapped them all some years before.

Wyeth discovered the area north of Fort Hall was still occupied by beaver, a geographic island that trappers hadn't unduly influenced. Wyeth described the country as having few buffalo, elk, and deer and only small numbers of mountain sheep, antelope, and rabbits. However, wolves serenaded them every night. There was only one readily available protein source that could provide reliable sustenance for wolves—beaver. For thousands of years prior to European settlement, the sheer number of beaver moving along and between stream corridors provided prey not only for wolves but grizzly and black bears, Canada lynx, bobcats, mountain lions, and the now-extinct carnivores from the Late Pleistocene.

Pelts destined either for the luxury clothing market or for the felting industries, of which hats were the most important, were the foundation of the trapping industry. Beaver hats in Europe became increasingly popular, eventually dominating the market until about 1830. The qualities of the beaver fur meant it held its shape in the rain, unlike the cheaper alternative of rabbit fur.

Making top hats was often lethal for hatters since mercury was used throughout the process of transforming beaver to felt. Prolonged exposure frequently led to mercury poisoning, with symptoms including early-onset dementia and irritability, muscular spasms and tremors, loss of hearing, eyesight, teeth, and nails. The mercury-poisoned mad hatter was immortalized in *Alice's Adventures in Wonderland*.

Between 1806 and 1838, beaver were brought to the brink of extinction from Missouri to California and from Canada to Mexico. Prices for beaver pelts were so high they were often referred to as "soft gold." Only in the 19th century did silk replace beaver in high-fashion men's hats. Had it not been for the silk top hat, beavers would have preceded the passenger pigeon into extinction.

Beaver have since recovered in North America to about 5–10 percent of their historical population. Most of the rebound in beaver numbers has taken place in areas without domestic livestock grazing. Normally, when beaver cut willows or aspen, the trees regenerate quickly with new shoots. Beaver were our natural equivalent of rotational farmers, leaving and then moving back into areas they had cut after 30 to 40 years. Livestock eat aspen and willow shoots that regenerate after beaver use an area. The resulting absence of aspen and willow ensures there is nothing for beaver seeking to return to eat or use for dams. This makes most streams on public lands in the western United States unsuitable for beaver recolonization.

While elk can have a similar effect; early indications in Yellowstone National Park are that beaver populations do have the ability to rebound in the presence of high elk densities as long as wolves are also present. Wolves prevent elk from congregating along streams and wetlands and overusing willows and aspen.

Because wolves that prey on domestic livestock are killed, beaver recovery isn't possible in western North America outside of national parks. Unlike elk, livestock are able to eliminate willows and aspen from the native plant community with impunity.

The largesse of historic beaver activity resulted in the creation of many of the most productive meadows and agriculture areas being used today. Beaver dams trapped sediment behind them. Over many centuries of beaver activity, valley bottoms built up deeper, more productive soils. The absence of beaver has resulted in increased soil erosion across western North America. These areas have become less productive, often precipitating the need for fertilizers. Without beaver, moist, productive meadows are continuing to be replaced with dry land vegetation.

While beaver are no longer in danger of extinction, they will never again shape the landscape in North America as they did for millennia. As beaver went, so did salmon. Beaver dams created ponds providing areas for young salmon to live, as well as insects and other food organisms for the fish to thrive. The ponds slowed stream flows and created quiet areas where the fish could rest and hide from predators.

Killing hundreds of thousands of beaver east of the Cascade Mountains was particularly harmful because of the more arid environment where salmon were living. The area is colder in the winter and warmer in the summer than coastal rivers. The beaver ponds helped modulate the temperature extremes in ponds and streams where salmon were present. The removal of beaver made it more difficult for young salmon to escape ice in the winter, hide from predators, and seek refuge from the heat in sheltered, shadowy areas in the summer.

The removal of beaver was the first significant blow to salmon populations. Since then, livestock overgrazing, logging, agricultural diversions, dams, commercial fishing, and warming ocean temperatures caused by global warming have all exacted an increasingly heavy toll. Today, salmon are extinct in almost 40 percent of the rivers where they were known to exist in California, Oregon, Washington, and Idaho. They are at risk of extinction in 44 percent of the streams where they remain. Estimates are that less than 0.1 percent of the tens of millions of native salmon still exist where they used to darken the rivers every summer and fall up and down the West Coast before white settlement.

When salmon spawn and die, completing their life cycle, their contribution to the stream ecosystem isn't done. The hundreds and thousands of salmon carcasses that existed historically would sustain and restore stream productivity, providing food for not only their offspring but insects, birds, and other species of fish. It was an ecosystem feedback loop that worked perfectly for millennia.

Forests continuously lose soil and nutrients to the water and ultimately the sea. Salmon coming back to their natal streams reverse this process. They eat fish and krill in the ocean, and their return produces a mass movement of nutrients from the ocean to the stream and forest floor. Historically, bears pulled hundreds of thousands of salmon onto the shores of rivers every year. Bears were nature's fertilizer spreaders, making forest soils more productive, ultimately enhancing the growth of trees and other vegetation. Fewer and fewer salmon returning from the ocean is creating increasingly sterile rivers, streams, and forests. The demise of salmon contributes not only to the downward spiral of their populations but countless other plants and animals that depended on the nutrient-cycling function they provided.

While forests need the salmon, likewise, salmon need the forests. The dense canopy of old-growth forests, aspen, and willow shades the stream from direct sunlight keeping the water cooler in the summer and warmer in the winter. Decaying vegetation supports populations of bacteria, fungi, and invertebrates that feed the baby salmon, called fry, as they develop. Fallen logs and branches in the stream provide obstacles that interrupt flows and provide resting areas for fish of all ages. Salmon and forests connect the air, the oceans, and even the hemispheres in a single, inter-dependent system.

CENTRAL NORTH AMERICA

As wind currents move east from the Pacific Ocean, more and more of the moisture drops out of the air on the western slopes of the Rocky Mountains as it pushes higher up to the summit of the range. This creates a metaphorical shadow of dryness on the plains east of the Rocky Mountains. The shortgrass prairie adapted to survive and prosper in the desert-like environment. By concentrating their roots at the top of the soil, and keeping their leafage minimal, grasses could conserve their moisture and take advantage of scarce rainfall. Their unique growth form enabled them to recover rapidly from drought.

Further east, the influence of the rain shadow waned and precipitation increased, giving rise to the tallgrass prairie. The root systems of tall grasses like big bluestem, Indiangrass, and switchgrass extended ten feet into the deep soils. In the areas between the eastern tallgrass prairie and western shortgrass plains, the grasses competed for dominance, the winner determined by available precipitation during the prior years. Unlike grasses in the Rocky Mountains and Pacific Northwest, those that evolved in the prairie were evolutionarily adapted to survive sporadic heavy grazing.

For thousands of years, the Great Plains was a relatively stable ecosystem prone to short-term extreme ecological oscillations depending on precipitation. This paradox provided for the long-term persistence of many plants and animals reliant on regular chaotic fluctuations in the environment. In the prairie regions of North America, two animals filled a void left by the departure of mastodons, giant bison, dire wolves, and other megafaunal species 10,000 to 20,000 years ago. At the time of the influx of European Americans, bison and the Rocky Mountain locust were the two keystone species in central North America. The effects of these two animals on the prairie were remarkably similar despite their extreme size difference.

Prior to the 17th century, bison numbers waxed and waned depending on the vagaries of precipitation and predation. Between 25 million and 30 million bison roamed the plains in North America prior to European settlement. In 1839, Thomas Farnham traveled through a herd of bison along the Santa Fe Trail for three days. He estimated the herd covered more than 1,000 square miles. Thirty years later, Major Richard Dodge encountered a bison herd 50 miles long and 25 miles wide with an estimated population of 4 million animals along the Arkansas River.

The European fashion statement that barely saved beaver from extinction had the opposite effect on bison. The advent of the silk hat in Europe ushered in an era where the American Fur Company and Hudson Bay Company switched from purchasing beaver pelts to bison hides. Using the fur for blankets, coats, saddles, and robes, Indigenous hunters were the first to fill the vacuum by trading bison hides to Europeans. By the end of the Civil War, European Americans had taken over the killing. Between 1870 and 1883, professional hunters butchered millions of bison, taking the hides and cutting out the tongues, leaving the carcasses to rot on the prairie. The slaughter that had begun in the 1830s became an unprecedented massacre. Federal authorities supported the hunting because they saw extermination of the bison as a means to force Indigenous tribes to submit to the reservation system. Thirty-one million bison were killed between 1868 and 1881. By 1889, there were fewer than 1,000 of these creatures left on the continent.

Historically, bison numbers were a model of consistency compared to the Rocky Mountain locust, which often fluctuated between incredible swarms and complete absence. The Rocky Mountain locust may have been the most abundant form of life ever to sweep across the planet.

A locust is a type of highly mobile grasshopper with the capacity to attain enormous population densities and a proclivity for aggregating and traveling in swarms. North America was blessed with a single species of locust. No known insect outbreak on the planet has ever approached the magnitude of the Rocky Mountain locust. A swarm of locusts in Nebraska in 1875 covered an area of 198,000 square miles and probably numbered 3.5 trillion individuals. Locusts naturally experience explosive population fluctuations. Outbreaks and crashes are triggered by their own population dynamics. A locust swarm had an apparent randomness that could shred one part of a county and leave the remainder untouched.

Swarms of these insects swept across the prairies, at one time reaching from southern Canada to the Mexican border and from California to Iowa. They were just as powerful a life force as the great herds of bison. So thick and massive were the swarms, they could turn noon into dusk. The last outbreak, 1874–1877, wiped out over half of the agricultural production in the United States west of the Mississippi River.

The locust shaped the cultural history of the western United States. So deeply rooted were these insects in the consciousness of Western culture that the European settlers of America were destined to interpret the Rocky Mountain locust infestations in profoundly religious terms. Locusts are mentioned in the Bible 36 times and the Quran twice, usually as a manifestation from God disciplining his people for some perceived injustice.

Pioneers and government agencies tried every imaginable method of control. They prayed for deliverance, organized bounty systems, conscripted able-bodied men into "grasshopper armies," and provided food aid to starving communities. Farmers tried to burn and beat the invaders,

drowning and plowing eggs or crushing and poisoning the hatching locusts, all to no avail.

While the Rocky Mountain locust disappeared as a dynamic phenomenon in 1875, the swarms continued to pummel America's heartland
into the 1880s, moving and settling with the whimsy of tornadoes. Their
devastation was like that of a living wildfire. Fifty to 100 tons of vegetation
per day were consumed by a typical swarm. Finally, in the 1890s, to the
relief of beleaguered farmers, the locust outbreaks subsided. When a small
swarm was reported in Manitoba in 1902, people wondered if another
period of devastation was at hand.

Nobody could have guessed that the Manitoba swarm would be the
last locusts seen in North America. Suddenly and mysteriously, the Rocky
Mountain locust disappeared. The last two specimens were collected by
a man in Manitoba, Canada, on July 19, 1902. In just 25 years, it went
from a biological marvel to having disappeared forever.

In an outbreak cycle, the locust swarms descended from the northern
Rockies in early June. These insects swept across the countryside, settling
wherever there was abundant food. As the locusts advanced to the south
and east, they began to mate, and soon after, the females began laying eggs.
The swarms left behind denuded grasslands riddled with eggs. The adult
locusts would live for perhaps a couple of months, seeding the countryside with the next generation. The embryos would mature through the
summer and then hibernate during the winter. The following spring, the
nymphs hatched on warm days forming into immense aggregations or
"bands." These immature locusts would march across the land, stripping
the vegetation to fuel their development.

This next generation would ride the winds further into the heartland
of the continent. After a three- or four-year buildup and an equal number
of generations, the outbreak would enter its final stage. Stretched to its
southern and eastern limits, a portion of the population would stream
back to the northwest in an apparent effort to return to its mountainous homeland, the Rocky Mountains. The locusts managed to endure

somewhere in the mountains, biding its time until conditions were again favorable for an eruption.

The locusts flourished during droughts when hot, dry weather weakened plant defenses and actually increased the nutritional value of the vegetation. The dry conditions also suppressed fungal diseases, which could devastate locust populations in wet years. The heat accelerated the locusts' maturation and development. In times of drought, lush vegetation was restricted to shady depressions that captured more moisture than surrounding areas. The locusts were forced to aggregate in these swales, a behavior that initially generated and then sustained the coherence of both nymphal bands and adult swarms. Outbreaks of the Rocky Mountain locust typically lasted three to five years.

It wasn't until nearly a century after the last documented locusts in Manitoba that Jeffrey Lockwood, a resourceful scientist at the University of Wyoming, solved the mystery surrounding their extinction. After nearly a decade of scientific detective work, Lockwood discovered the locust had been extremely vulnerable to even small-scale human disturbances during periods between outbreaks. The fertile river valleys of the mountainous west represented a sanctuary where the locust had always found what was necessary to persist in the face of adversity.

Small geographic areas that sustain populations are termed ecological bottlenecks. A population of any plant or animal is only as safe as its weakest link. An ecological bottleneck can spell disaster for a species if the compression of its numbers occurs in a time and place where human disturbance is likely to occur. Farming and domestic livestock grazing in the valleys near streams and rivers in the Rocky Mountains were primary contributors to the demise of the locust.

Between 1870 and 1884, the number of cattle in the western states grew from about 450,000 to nearly 40 million. Farms and grazing by cattle and sheep were heavily concentrated along rivers and streams, precisely the habitats utilized by the locusts. Domestic livestock aren't mobile and need plentiful water. The aggregation of cattle in the valleys near rivers and streams dramatically altered these ecosystems, not only negatively impacting salmon populations but obliterating the locust. The extinction of the Rocky Mountain locust was the result of unplanned, uncoordinated, and unknowing human activities in concert with the normal vagaries of nature.

The synchronicity of drought and locust outbreaks significantly reduced available grasses in the prairie magnifying the periodic contraction of bison numbers. The temporary absence of both bison and locust for periods of time following droughts provided an opportunity for the grasslands to regenerate and species to reestablish.

The locust and bison collectively did the same thing in the prairies the passenger pigeon and beaver did in the eastern and western portions of North America. Their propensity for aggregating in either large herds or swarms and occasional absence created a diversity of grassland habitat in both form and function. The mosaic of grazed and ungrazed areas provided habitat for prairie chickens, prairie dogs, burrowing owls, swift foxes, and numerous grassland birds and small mammals.

Nearly every organism on the plains was significantly affected in some manner by the loss of bison and the locust. There were probably several lesser-known, low-density, difficult-to-detect plant and animal species that slid into extinction in the aftermath of the bison and locust's collapse, completely unknown to humanity. Even native fish species were affected by their absence. Because of the impact on migratory birds and bats, and the normal ebb and flow of energy, carbon, and nitrogen across the plains and between continents, the ecological effect of their demise was felt throughout the entire Western Hemisphere.

The locust outbreaks and increases in bison numbers would result in dramatic population increases for those animals utilizing them for food. As the locusts disappeared and bison numbers contracted, gradual or precipitous declines in dependent wildlife numbers occurred until the next outbreak when they would get another infusion of nutrients. The entire prairie evolved under this cyclical, chaotic influence.

The regular use of fire by the earliest human prairie inhabitants precipitated ecological changes, likely reducing the extent of forested areas
and expanding grasslands. This benefited Indigenous people by making
the landscape more suitable for bison. They derived food, clothing, tent
material, and tools from hunting bison.

Horses first became available to the Plains Indians as a result of their
diffusion northward from the Spanish colonies in New Mexico in the 17th
century. This historical event precipitated the downward population spiral
of both Indigenous tribes and bison in North America. On horseback,
Native hunters could follow the migrating herds more closely and over a
wider range, kill the animals more efficiently, and carry back more meat and
hides. Horses also directly competed with the bison for forage and water.

The ease of hunting bison using horses generated a magnet for
Indigenous tribes in adjacent areas. In a single century, it created one of
the most renowned hunting cultures in history. During this era, some
tribes completely reinvented themselves from hunter-gatherers to highly
specialized equestrian nomads on the high plains. Both the emergence
of these tribes in the 18th century and their disappearance in the 19th
century were directly related to the European invasion of North America.

First, domestic livestock and then the plow successfully eliminated
the eastern tallgrass prairie. At one time, this ecosystem covered portions
of 14 states and enveloped an area approximately the size of Texas. The
shortgrass prairie was too dry and unproductive to farm economically
although it was not for a lack of effort by early settlers. Failed farming
incursions along with unprecedented domestic livestock grazing resulted
in a significant loss of topsoil and degradation of large expanses of the
shortgrass prairie.

Grazing by bison favored the dominance of shortgrass immediately
east of the Rocky Mountains. The disappearance of the bison led to aspen
invading a broad belt of the Canadian grasslands, the spread of mesquite
in the southern plains, and allowed the invasion of other plant species all
across the western plains.

The Homestead Act of 1862 spelled the end of the tallgrass prairie. The promise of free land and a new start created a tremendous boom that drew homesteaders West. Any head of household who was at least 21 years old could claim a 160-acre parcel of federal land. In 1871, there were more than 20,000 homestead entries staking claim to 2.5 million acres of land. Nearly 2 million people moved to the western prairies and mountains in the 1870s. In 1889, the Oklahoma Land Rush resulted in additional claims on more than 2 million acres of land. Entire towns of 10,000 inhabitants were created in a single day. Between 1870 and 1900, as much land was settled and brought into agricultural production as had been transformed in the previous 250 years of North America's history. A grassland ecosystem shaped by thousands of years of climate, geology, and evolution disappeared in less than half a century.

FORGOTTEN LESSONS

The historic numbers of passenger pigeons, beaver, Rocky Mountain locust, bison, and salmon are beyond our ability as humans to appreciate or comprehend. While attending work meetings, I had the opportunity to watch the colony of Mexican free-tailed bats exit the Congress Avenue Bridge spanning Lady Bird Lake in downtown Austin, Texas. It was winter, so there were only a quarter of a million bats using the bridge. Despite that, the outflight of bats in the evening was an incredible spectacle.

During the half hour they emerged, they created a virtual aerial river about 150 yards wide for as far as the eye could see. It was something a person needs to view to believe. Flocks of passenger pigeons were an estimated mile wide and took three days to pass overhead. As incredible as the spectacle seemed, the bat exodus from Austin's Congress Avenue Bridge was a drop in the proverbial bucket in comparison to passenger pigeon flocks.

As a result of the demise of the passenger pigeon, in 1900, Republican Congressman John F. Lacey of Iowa introduced the nation's first wildlife protection law. Lacey said on the House floor, "We have given an awful exhibition of slaughter and destruction, which may serve as a warning to all mankind. Let us now give an example of wise conservation of what remains of the gifts of nature." Also inspired by the extinction of the passenger pigeon, the Migratory Bird Treaty Act in 1918 was passed protecting not just birds but also their eggs, nests, and feathers.

Sadly, 100 years later, attempts were made to weaken the Migratory Bird Treaty Act so industries are no longer accountable for "incidental" bird deaths. A century of migratory bird conservation efforts passed because of the passenger pigeon's extinction was nearly unraveled on a political whim. The warning to all humankind that Congressman Lacey spoke of is long forgotten. The amnesia of humanity toward our environment is the stuff of legend.

Over the last four decades, there have been numerous attempts to undermine the Endangered Species Act, the landmark environmental legislation of not just the United States, but the entire planet. While some politicians may be the face of anti-environmentalism, their actions are at the behest of the American people. No matter how deep humanity wanders into the metaphorical coal mine, no price is too steep or warning too dire to usurp human growth and prosperity.

Even if we had managed to conserve the last of the Rocky Mountain locust or passenger pigeons in a zoo, they would be no more their original species than the condors that can never again know the vast, unbroken expanses of land in the California foothills.

The extinctions in North America during this period of time were not unique and represent a subgroup of what was occurring around the planet. The Portuguese ibex went extinct in 1892. Australia lost a host of birds and mammals between 1850 and 1930 including the Tasmanian tiger. The Indian Javan rhinoceros in Asia was gone by 1925. All of these human impacts to biodiversity and native ecosystems predate bulldozers and chainsaws but set the stage for subsequent years.

Human arrival in the Caribbean around 6,000 years ago correlated with the extinction of many species including ground and arboreal sloths. Fifteen hundred years ago, a flightless bird 10 feet tall and weighing almost 1,000 pounds—and the giant lemurs, the globe's largest primates—disappeared precisely when the first humans set foot on Madagascar. Similar ecological disasters occurred on almost every one of the thousands of islands in the Pacific Ocean, Indian Ocean, Arctic Ocean, and Mediterranean Sea. The extinction of the largest eagle that ever existed, the Haast's eagle with a nine-foot wingspan and their main prey the giant moa, occurred in the 15th century in New Zealand. The great auk was a flightless bird similar to a penguin and at one time numbered in the millions. The last known great auks were slaughtered in 1844. *Homo sapiens* is singularly responsible for an extinction continuum spanning 2 million years.

Two of the most powerful life forces in the history of complex life on Earth were victimized by the human fallacy that no amount of exploitation could endanger creatures so abundant. The passenger pigeon and Rocky Mountain locust were destined for a relatively quick extinction subsequent to the first Europeans arrival in North America. Their tipping point, however, was the same as ours—nearly 2 million years ago. Their arrival at the inevitable end of existence was only a few centuries earlier than ours will be. They did manage to outlast many plants and animals that disappeared even earlier by inhabiting a continent further removed from the planet's earliest hub of human habitation.

Entire ecosystems have now vanished. The bison, beaver, American chestnut, and salmon have now been reduced to mere husks of their former selves. None of these organisms were obscure canaries in a mine. They were pterodactyls in our living room. Few people now even know of their existence or the prominent role they played in shaping landscapes. The warnings ignored and forgotten have now transcended centuries and millennia.

The fragility of nature has always been beyond our comprehension. The distress calls of other organisms on the planet are increasingly loud and more frequent. Our ancestors were exuberant contributors as one species after another winked out. We're now a combination of both active participants and unimpressed spectators as the dire, imminent warnings spool across the internet and television daily. We've evolved from a creature capable of never-before-seen annihilation and cruelty to one now hitting the self-destruct button with a child-like enthusiasm. The inexorable ship to humanity's exodus from the planet set sail thousands of years ago.

CHAPTER 4

ATTITUDE

The attitudes people exhibit toward the planet's natural resources consistently follow two themes. The first is: If there is a lot, take a lot; if there is a little, take it all. The second: If we don't use natural resources, they will go to waste. These schools of thought repeated themselves over and over throughout my career. While the characters and props were incredibly diverse, the plot never changed.

In a review of the actions of *Homo sapiens* over the last 50,000 years, the same concepts that I experienced and observed throughout my career continually repeated themselves. Human behavior toward the use of our environment hasn't changed in thousands of years.

LIVESTOCK GRAZING

Country-western music, western movies, and Marlboro commercials have left Americans with starry-eyed visions of dusty cowboys on horseback crooning to cattle scattered across beautiful landscapes. Behind this idealized vision is a nasty reality of public lands abuse, out-of-control subsidies, and a victim narrative based on an infantile sense of privilege.

Having been raised on a small ranch with close relatives who were also ranchers, I was one of the many naively idealistic Americans who thought ranchers were the cat's meow. While working for land management agencies, I was quickly bludgeoned with a reality that in no way matched the perception acquired in the early part of my life. The first concept ranchers happily introduced me to was that if there was a lot of grasses and vegetation for their livestock, stockmen took a lot; if there was a little, they took it all.

Battles over who should be able to graze livestock on the millions of acres of public land in the western states have raged for over 150 years. Livestock ranching there began growing into a major industry following a brucellosis outbreak in Great Britain and the killing of southern cattle by the Union Army during the Civil War. New railroads helped transport cattle to markets, and ranching costs became relatively low. Many early ranchers started leaving cattle to roam without boundaries or fences, a practice known as open range grazing. Stock growers decided among themselves who was allowed to use which land.

Unregulated grazing was an abysmal failure and resulted in numerous range wars across the West. These conflicts resulted in the death of thousands of people in the late 1800s. Conflict between large cattle ranchers and homesteaders, disagreements between ranchers over water rights, and disputes between sheep ranchers and cattlemen all played a part in the rangeland free-for-all. Range wars occurred in almost every western state, but Wyoming provides a well-documented example. Between 1870 and 1884, cattle numbers in the state increased from 8,000 to more than 1 million. Lasting from 1889 to 1893, the Johnson County War has in the modern era provided fodder for numerous popular novels, films, and television shows.

In the early days, land and water rights were usually distributed to whomever settled the property first, and farmers and ranchers had to respect those boundaries. As more and more homesteaders moved into Wyoming, competition for land and water soon enveloped the state. Cattle companies reacted by monopolizing large areas of the open range to prevent newcomers from using it. The large ranching outfits in Wyoming organized as the Wyoming Stock Growers Association and held a tremendous amount of political sway in the region.

Well-armed outfits of horse and cattle rustlers roamed across portions of Wyoming and Montana in the 1880s. Using the idea of the rustlers to their own benefit, the Wyoming Stock Growers Association targeted any homesteaders perceived as making inroads into what they considered as their grazing territory. The association employed detectives to eliminate the supposed criminals—labeling settlers as rustlers was routinely used to rationalize barbarous actions against them. As the competition grew more intense for grazing land and water, large cattle companies began forcing settlers off their land and setting fire to their properties.

On July 20, 1889, a detective from the association accused a Johnson County rancher of stealing cattle from a fellow stockman. The cattlemen sent riders to seize the rancher and her husband and lynched them both. Hanging a woman appalled many of the locals. The county sheriff arrested six men for the lynching, and a trial date was set. Three of the witnesses who would testify against the aggressors either mysteriously disappeared under unknown circumstances before the trial or were known to have been murdered.

Enemies of the Wyoming Stock Growers Association soon fought back by killing the range detective who accused the couple of stealing cattle. A series of murders on both sides escalated the conflict. It also led to a local cowboy named Nate Champion forming a competing Northern Wyoming Farmers and Stock Grower Association to challenge the Wyoming Stock Growers Association.

Prominent cattlemen in the Wyoming Stock Growers Association sent an assassination squad of three men to kill Champion on November 1, 1891. The assassination went awry when one of their crew was killed and the others were sent fleeing. Champion, at least temporarily, escaped unscathed, but in the months after the attempt to murder him, the violence continued to escalate.

Unfazed and ever more determined by their failure to eliminate Champion, the Wyoming Stock Growers Association sent agents to Texas to recruit gunmen. The intention was not only to kill Champion but to break up the newly formed competition group. The leader of the hired gunmen for the Wyoming Stock Growers Association had a list of 70 county residents to be either shot or hanged and a contract to pay the Texans five dollars a day plus a bonus of $50 for every rustler, real or alleged, they killed. At some point, the Texans ambushed and killed nine trappers, whom they mistook for rustlers, and received a $450 bonus.

Their priority target was always Champion, whom they succeeded in killing in a gun battle at his ranch in April 1892. Another settler happened by during the ambush, and the invaders gave chase, shooting at him. He managed to escape back to Buffalo, Wyoming, where he reported the attack on Champion to the townsfolk. The sheriff in Buffalo raised a posse of 200 men and set out for Champion's ranch the next day.

News of a major hostile force heading for them reached the Wyoming Stock Growers Association's hired guns. They sought protection by holing up in a log barn on a remote ranch. A number of fighters on both sides were killed in a battle that lasted two days. The governor of Wyoming intervened on behalf of the Wyoming Stock Growers Association's agents, sending a telegram to President Benjamin Harrison for help. Harrison immediately ordered the Sixth Cavalry from Fort McKinney to the ranch. The commander negotiated with the sheriff an end to the siege by agreeing to turn the invaders over to civilian authorities. The hired guns were taken to Cheyenne to await trial.

The evidence was said to implicate more than 20 prominent stockmen in Cheyenne as well as men in high authority in the state of Wyoming. The invaders were protected by a friendly judicial system and took advantage of the cattle barons' corruption. Charges were never filed against the men in high authority, and the invaders were released on bail. Most fled back to Texas and were never seen again.

After the men's release, tensions remained high in Johnson County with more ranchers, settlers, and soldiers killed on both sides of the conflict. Emotions ran high for many years. Afterward, a number of tall tales were spun by both sides in an attempt to make their actions appear morally justified. Many viewed the rescue of the hired killers, at the order of the

Republican President Harrison, as a serious political scandal with overtones of class war. As a result, the Democratic Party became popular in Wyoming for several years.

In 1897, Wyoming Governor William Richards knew overgrazing was rampant and the conflicts never-ending. He suggested to the legislature that if the state could control the public range and lease it at low rates to grazers, overgrazing would be reduced, taxes would fall, and range conflicts would end. However, the idea stalled. It wasn't until the following decade during the conservation-minded Theodore Roosevelt administration that grazing leases were instituted.

Forest reserves, which would later become national forests, were established in the West by the Forest Reserve Act. Grazing was officially excluded from forest reserves three years later, but the lack of enforcement made the law a mere paper tiger. The first head of the United States Forest Service, Gifford Pinchot, later advocated for and prevailed in establishing a regulated grazing system. The first grazing fee system was established in the forest reserves in 1905. Many stockmen called it unconstitutional, but the Supreme Court ruled against them.

In 1907, representatives of Western ranching and mining interests initially called for public lands to be transferred from federal control to the states. However, President Theodore Roosevelt prevailed in maintaining federal authority. Even then, many ranchers acknowledged the widespread overgrazing and need for reforms. Others claimed the problems had natural causes and were not the result of overgrazing. They insisted it was a drought and when rainfall once again normalized, the range would quickly recover.

Despite those claiming the problems were natural, it was generally accepted that large areas of public lands had already suffered serious vegetative damage and accelerated erosion, primarily as a result of uncontrolled grazing. To deal with the ubiquitous overgrazing and mounting conflicts among livestock operators, in 1934, President Franklin Roosevelt signed the Taylor Grazing Act. It laid a foundation for the West's modern grazing system. Even some of the largest ranchers within the Wyoming Stock

Growers Association supported the legislation because it put an end to homesteading.

Just days after passage of the Taylor Grazing Act in June 1934, Henry A. Wallace, the Secretary of the US Department of Agriculture, was the keynote speaker at the Wyoming Stock Growers Association. In an all-too-typical adolescent effort at bullying and intimidation, longtime Johnson County rancher Elmer Brock rapped the butt of a six-shooter on the podium when calling the meeting to order.

Undeterred by the charged atmosphere, Wallace told the ranchers, "It is all right to go ahead if you want to under your rugged individualism and overstock your ranges and eat off your good pastures; it is all right for you to hurt yourselves if you want to; but it is a shame to hurt the land the way you have been doing."

Passage of the Taylor Grazing Act resulted in the creation of allotments, assigned areas for stockmen to graze their livestock. While the act's passage ended disputes over whose livestock could graze which areas, it began disputes over whether grazing permits should be treated as private property. Neither the Taylor Grazing Act nor any other federal laws describe grazing on public lands as a right. However, the term *grazing rights* became popular with the livestock industry, and it is still used today—their apparent theory being that repeating it frequently and for extended periods will somehow make it come true.

In 1946, the General Land Office and the Grazing Service merged to create the Bureau of Land Management (BLM) under the Department of the Interior. The BLM would come to manage more land than any other agency—245 million surface acres and 700 million acres of subsurface mineral estate. In 1966, a grazing fee for a cow and calf for a month on public land was established at $1.23, to be adjusted annually according to a formula created by livestock industry advocates.

To illustrate just how rancher-friendly the formula, the grazing fee in 2018 was $1.41. If the standard rate of inflation was applied to the 1966 grazing fee, it would be $9.37 today rather than $1.41. This is made possible by Americans convinced that all taxes are theft, but subsidized overgrazing, increased erosion, and deteriorating fish and wildlife habitat on public lands is not.

The expression "don't look a gift horse in the mouth" was lost on public land ranchers; in 1970 Pankey Land and Cattle Company in New Mexico led a class action lawsuit against the secretaries of agriculture and interior, alleging that the federal government did not have the authority to increase fees. The case was dismissed.

In 1976, under Republican President Gerald Ford, the Federal Land Policy and Management Act was passed, creating a mandate for the BLM to manage its lands for not only livestock grazing but fish, wildlife, and other uses. However, changing the fundamental mission of an agency with a half century of institutional livestock advocacy has never happened. The BLM is still euphemistically known as the Bureau of Livestock Management.

Because of the agency's cozy relationship not only with the livestock industry but more recently with oil and gas developers, it is becoming better known for managing public lands for multiple abuses. One of the more egregious recent examples involved the BLM manager of the Farmington, New Mexico, office, who was found to have accepted gifts and meals from the oil and gas industry, failed to financially disclose them, and failed to act impartially in the performance of his official duties. He was rewarded for his indiscretions with retirement from the BLM without penalty and a subsequent new role as president of the New Mexico Oil and Gas Association, where his annual salary was more than $250,000.

Early in my career, the phrase "if there is a lot, take a lot; if there is a little, take it all" started out as a joke in describing the actions of those public land ranchers we dealt with on a regular basis. Witnessing continual overgrazing and desertification in process quickly wrung any humor out of our little wordplay.

Like all plants, grasses harness energy from sunlight and turn it into chemical energy they use to grow, flower, and set seed. The leaves of plants contain cells called chloroplasts which capture energy from the sunlight.

Without their leaves, plants can't survive for long. Native grasses are no different than house plants. If 10 percent of the leaf surface is removed, the plant will almost certainly recover, but all their energy will be transferred from growing and reproducing to recovering. Given enough time and nurturing, a plant might still be able to recover after the loss of 20 percent or even 30 percent of its leaf surface. However, ranchers on public lands have routinely allowed their livestock to take between 50 and 100 percent of the leaves of native grasses and flowers year in and year out over vast areas.

Grazing that is too close and too frequent makes it impossible for the plants to recover and allows less productive invasive plants to establish. The healthy native perennial grasses are gradually replaced by sod-forming grasses that store more of their nutrients in root systems or annual grasses whose availability to livestock and wildlife is short-lived and unreliable. Over time, invasive grasses that replace the native grasses are those that can still survive continual excessive grazing. As a result, grasslands have been converted from native grasses and flowers that were between 12 and 36 inches tall to those that barely peek above the surface of the soil and are capable of weathering the continual livestock maelstrom.

In years where the precipitation was normal or above average and growing conditions for plants optimal, the ranchers pushed to get their cows on public rangelands early in the season. The cows remained until they ate as much of the available vegetation as possible before returning to the private lands in the fall. During drought conditions, they did not adjust their timelines to compensate for less vegetation resulting in nothing but dust over large areas. After eating everything they could, the livestock stomped most of what was left into dirt in efforts to satisfy their hunger. The areas closest to water were the most seriously impacted. The only places that weren't annihilated were those so far away from water that the cows couldn't physically get there and back before being forced to turn around by thirst.

I was repeatedly told that any problems with the deteriorating rangeland conditions were due to natural causes and were not the result of overgrazing. Echoing the same argument used by their predecessors, ranchers insisted it was drought and when more normal amounts of rain fell, the range would quickly recover.

Many of them sincerely believed this was the case. When the rains did come, the hills would turn green with new plant growth. The distinction between productive native grasses and flowers or invasive, unproductive grasses and exotic weeds was lost on them. One could paint the dirt green, and as long as the ranchers didn't get too close, they would have considered their range recovered.

In the college range management classes I attended, the areas in the immediate vicinity of stock-watering tanks and fences were described as sacrifice areas. These were the areas that were normally expected to be tromped into dirt by livestock as a result of their frequent use. In class, the definition of "immediate vicinity" meant a 10-to-20-yard radius around a water trough. The reality was that sacrifice areas on public lands extended from a quarter to a half mile from any water source. With enough water troughs scattered around the landscape, the grasses and flowers on large areas of public land were completely obliterated.

At one time, there were thousands of natural springs where water bubbled out onto the surface of the landscape from groundwater fueled by snowmelt. These springs were scattered randomly across western rangelands. They were the veritable oases in the desert, providing succulent vegetation and clean, dependable water for wildlife in very localized areas.

One by one, natural springs have been dug out with a backhoe and culverts placed where the water surfaces, piping it into stock tanks for livestock. The goal was getting water to every conceivable location so livestock could eat as much of the grasses in the stockman's allotted area as possible. Many of the springs in dry land, non-forested areas in the western United States have been developed in this manner. Destroying these native spring systems has likely resulted in the extinction of hundreds of aquatic invertebrate species including springsnails, amphipods, isopods, insects, and other snails that were never identified before being ushered out of existence.

Whether it was water from springs or native grasses and flowers, if there was a lot, the ranchers took a lot, and when there was only a little, they took it all. The same thing continues to occur with more native grasslands being converted into unproductive expanses of annual grass and weeds every year.

Hiring squads of assassins to kill those who might interfere with their ability to graze livestock on public lands is no longer copacetic. However, the political power western cattlemen wielded a century ago still exists today. They very capably handcuff efforts that might interfere with overstocking western public rangelands. While normally less overt than what transpired at the 1934 meeting attended by Henry Wallace, more subtle forms of bullying and intimidation are utilized to keep recalcitrant federal employees in line.

Whenever serious overgrazing problems came to light, it invariably led to a tour of the allotment problem. The rancher who grazed livestock on the allotment would invite friends and cronies sympathetic to his cause to attend the tour. It usually included several of his colleagues, a county commissioner, a range extension specialist, and a state representative. The agency participants included the range management specialist, the area manager, and either the fisheries specialist or the wildlife biologist.

The typical tour would stop at some overgrazed locale where everyone would get out of vehicles or off horses and kick rocks while the rancher told everyone how well things were going, completely ignoring the obvious overgrazing. All the livestock lackeys would sagely nod their approval and suggest ways to pipe water to some obscure spot that livestock hadn't yet destroyed, furthering the degradation.

In most cases, the federal agency manager had mastered the art of incessant yapping without saying anything. I was usually quietly embarrassed for them, although the ranchers didn't care as long as the manager continued to support their ability to overgraze the rangeland. After all the rocks had been kicked and self-congratulations delivered, someone would say, "If no one has any issues, let's move on to the next site."

For the fish or wildlife biologist on the tour, these were career-defining moments. Say nothing, and they are everyone's friend. The rancher has found a biologist who will help them rationalize their overgrazing to outside interests. The manager has an employee who hasn't caused any conflict, potentially endangering their quest for moving up the bureaucratic ladder. As long as the biologist consistently says and does nothing, they too will make themselves eligible for future promotions.

In natural resource management, employees who can consistently avoid conflict and do nothing can aspire to future advancement in the organization. As long as the person has no resource ethic, it's absolutely the greatest job in the world. Everyone is your friend, promotion is easy, there are no conflicts, and minimal work is required.

If, however, the biologist uttered the dreaded phrase, "There is an overgrazing problem here," the rancher and all his cronies would immediately get angry and begin attacking their credibility. Being verbally assaulted by a gang of people in a remote area whom you will be spending the remainder of the day with is never a pleasant experience. If they didn't say anything because the situation was too intimidating but then raised the issue later, their consternation would be dismissed because they had an opportunity to air their concerns on the tour. The desired outcome of the tour for nearly everyone was affirmation of the status quo by garnering an intimidated silence from the attending biologist.

Articulating a candid assessment of overgrazing by the biologist was the classic lose-lose situation. The agency manager was never happy to have controversy of any form. It was bad for their aspirations of promotion to ever higher positions within the agency because they were perceived as unable to control their employee. The employee would invariably get a poor evaluation for the year because they couldn't get along with the public.

Most of the time, one of these unhappy moments was all it took to ensure the employee never again said anything that would anger anyone. They chose either a resentful silence or finding another field and eventually leaving the agency. Normally, a little bullying and intimidation solved the problem of an ethical employee in perpetuity.

Despite being a pacifist at heart, I found I could not remain silent about the deteriorating conditions of public rangelands. It was too personal for me. Regular verbal abuse and intimidation from participants on range tours came with pointing out the obvious. I was the focus of ire on multiple rangeland tours, some with as many as 50 or 60 ranchers in attendance. On one occasion while I was describing why overgrazing areas along streams was particularly harmful, one rancher "accidentally" lost control of his horse, knocking me down.

When emotions ran especially high, the attacks usually went from professional to personal. There was a tremendous amount of frustration both within the local livestock industry and the agency itself when intimidation didn't result in the silence of participating agency environmental specialists. Not all ranchers had the self-control to disguise or keep their intimidation subtle. Twice I attended meetings with a rancher when he threatened to punch a female environmental specialist half his size.

Livestock operators and agency managers made environmental candor from agency staff predictably unpleasant. Even after convincing managers with definitive evidence that overgrazing was the problem, not a lack of rain, the stockmen used elk as their scapegoat, unduly influencing managers in the process. It wasn't the stockmen's cows' fault—there were too many elk. The stockmen's overzealous lust for cheap forage at the public's expense gave new meaning to the term *sacred cow*. While Hindus worship the animal, American stockmen worship the money derived from their sacred cows.

There was a small minority of ranchers who were sincerely trying to treat the vegetation and resources with care. I both admired and respected their attitudes. Unfortunately, they were trying to fit a square peg in a round hole.

The dryland areas in the West evolved under almost no grazing. Research showed that bluebunch wheatgrass, one of the most prevalent native grasses in the West, took seven years to recover after being grazed once in the spring. Any grazing had deleterious effects that only multiple years of complete recovery could counteract. Trapping journals corroborated what the physiology of the grasses clearly showed. Bison, elk, and deer were uncommon in the Rocky Mountains and Great Basin of the West. Bison and elk, in particular, were Great Plains animals. Elk and deer are now found almost exclusively in the Rocky Mountains and Pacific Northwest in numbers and densities far exceeding natural levels.

Making an already impossible problem worse was the century of overgrazing that preceded these more ethical ranchers trying to treat the land with care. There is no grazing that doesn't further exacerbate an already deteriorated ecosystem.

For 3.7 billion years, our planet's natural resources have been going to waste. Only in the last 10,000 years have humans arrived on the scene to correct this horrible injustice. My introduction to this concept was once again compliments of the majority of public land ranchers. "Look at all that feed just going to waste" was one of the most common utterances I heard while working with or around ranchers. It was their mantra. I quite literally heard it hundreds of times.

A grass still capable of photosynthesis was a personal affront and challenge to the stockman. Politically powerful ranchers pressured public land managers to extend pipelines for miles or even haul water to keep those damnable grasses from doing what they had done for millennia. No taxpayer expense was too great to get that last blade of grass into their cows' digestive tract. Millions of dollars of taxpayer-funded livestock projects were well worth their investment in whine as they pocketed a little extra money at the feedlot for slightly heavier cattle.

WATER

The high mountains of southern Colorado are home to the headwaters of the Rio Grande, the third-longest river in the United States. The river runs south through New Mexico before turning southeast. It runs along the southern edge of Texas, serving as the border between the United States and Mexico, since 1848, before emptying into the Gulf of Mexico. Will Rogers once described the Rio Grande as "the only river I know of that is in need of irrigating."

Sixteenth-century Spaniards called the river Rio de las Palmas. The bright forest of palm trees around the mouth was a landmark for navigators of the Gulf of Mexico. In good years, the lowlands surrounding those groves would be marshes teeming with shellfish and minnows being hunted by ibis, herons, speckled trout, and red drum.

The sprawling river delta has now been reduced to a nearly barren, eroded strip of earth, and some residents of Port Isabel are having trouble breathing because there is so much wind-blown grit in the air. A sandbar at the mouth of the Rio Grande has left the mythic river a tepid, stagnant shallow having too little water to cross the bar and reach the ocean.

Bosque is a Spanish word that refers to a unique ecosystem created by the Rio Grande consisting of woodlands dominated by cottonwood trees and willow thickets connected to the ancient river running through them. The Rio Grande bosque once consisted of forest patches interspersed with wet meadows, marshes, sloughs, ponds, and small lakes. The bosque in the Middle Rio Grande Valley supported a rich assortment of plant and animal life. Prior to Spanish settlement, the bosque in some places was nearly a mile wide. Due to water withdrawals from the river for a myriad of uses and urban encroachment, the current-day Rio Grande bosque is a mere shell of its former self. The population of mature cottonwood trees is nearing the end of their natural life with few young seedlings. The Rio Grande bosque is itself an endangered ecosystem because without the river, there can be no bosque.

The history of irrigation in the Rio Grande valley predates Columbus. Ancestral Puebloans are best known for their stone-and-earth dwellings built along cliff walls accessible only by rope or rock climbing. However, they may have practiced some forms of agriculture in the 11th and 12th centuries. Irrigation agriculture is well-documented around 1590 after some Pueblos were missionized by the Spaniards. With the colonization of New Mexico by the Spaniards around 1600, irrigation became increasingly prevalent.

By the 1890s, water use in the upper Rio Grande Basin was so great that the river's flow near El Paso, Texas, was reduced to a trickle. The Rio Grande Basin now has a complex system of dams and canals for storing and diverting water for irrigation. The first year in history the Rio Grande did not reach the Gulf of Mexico was 2001. The Rio Grande was predeceased 38 years by its larger sibling, the Colorado River, which first failed to reach the Sea of Cortez in 1963.

Dams and diversions have already resulted in the extinction of two minnow species in the Rio Grande, and one other has been barely clinging to existence for the last 30 years. The appropriately named Rio Grande silvery minnow is critically endangered. The minnow was historically the most abundant fish in the Rio Grande Basin and was found from 55 miles north of Albuquerque to the Gulf of Mexico. Now it survives only in a section of the river near Albuquerque.

Upon my transfer to Albuquerque in 2000, irrigators and conservationists were embroiled in a legal battle that would determine the immediate fate of the silvery minnow and the river. At the time, the US Fish and Wildlife Service insisted on a continuously flowing river to prevent the fish's extinction. However, by my second year in New Mexico, the Clinton administration was history and the Bush administration asserted its political will.

The agency was forced to change its story to fit the new political reality. The new biological story had to match the fantasies of politicians and irrigators. Diverting all the water from portions of the river where the silvery minnow still existed became biologically acceptable. "Non-water" solutions to conserve the Rio Grande silvery minnow became the buzzword. Because biology doesn't change with the whims of bureaucrats and politicians, conjuring and subterfuge normally the purview of great magicians became necessary.

After over 20 years in the bureaucracy, I found it a scenario that had become all too familiar. Biologists were asked to saddle up our unicorns and start chasing rainbows at the behest of the administration, agency managers, and politicians. Those successful in giving the illusion of catching the rainbow moved up the agency management ladder to positions of greater responsibility. There they spun facts into even more colorful rainbow illusions. Because the new story didn't fit reality, it was necessary to change reality to fit their story.

The whole premise for conserving the Rio Grande silvery minnow was analogous to coming up with "non-air" solutions for humans. The non-water solutions allowed the irrigators and Bureau of Reclamation to dry up portions of the few areas where the silvery minnow still existed.

Captive propagation was the term used for the non-water fantasies requiring capturing silvery minnow in the drying river as irrigators divert the water. Those fish that survive the procedure are taken to a hatchery. When there is enough water in the river for them to survive, they are reintroduced back into the river. It would be similar to humans leap-frogging between the middle of a wilderness to downtown Tokyo, Japan and back again.

Fish mortalities occur going, coming, and during their transition to the new environment. In addition to its obvious biological absurdity, the exercise ignores the stated purpose of the Endangered Species Act, namely to "provide a means whereby the ecosystems upon which endangered species and threatened species depend may be conserved."

During my tenure in Albuquerque, absurdity became the norm. The pressure on biologists to conform to the agency's knee-jerk reactions and the ever-changing political climate were intense. Agency personnel pointing out the ridiculous nature of the proposals were quickly bullied into silence.

In the Southwest, water is gold. Users can never get enough. In years where snowpack and monsoon rains provide plenty of water, irrigators use a lot of water. During drought years when little water is available, irrigators use it all. Water policy in the region consists of crossing fingers and hoping there won't be a drought. During my tenure as endangered species branch chief, irrigators, the Army Corps of Engineers, and the Bureau of Reclamation rode the ragged line of providing barely enough water that the silvery minnow could eke out another year of existence, but no more. The concept of conserving water in years with surplus precipitation for those years when it was scarce was never a consideration.

Those vying for every last drop of the Rio Grande are numerous and politically powerful. Texas, Colorado, New Mexico, and Mexico all want their share of the Rio Grande. Albuquerque, Santa Fe, Las Cruces, El Paso, McAllen, Laredo, Brownsville, and Ciudad Juarez all want their share. The International Boundary and Water Commission, Interstate Stream Commission, and Rio Grande Compact are all agencies and institutions that squeeze the Rio Grande for their constituents and purposes. Irrigation Districts lining the Rio Grande have their own politically powerful farmers trying to maximize diversions from the Rio Grande for their crops. The once-mighty river is now the mouse wandering through this nest of vipers.

Many of the entities battling for the water would like nothing better than to funnel all of it into a pipeline or ditch so it isn't "wasted" on fish, wildlife, plants, and a free-flowing river. The water not being frittered away on the river even has a name—working water. It's the water that is making people money as opposed to the squandered water that has for time immemorial made its journey to the ocean.

Diversions for municipal and agricultural use already claim more than 95 percent of the Rio Grande's average annual flow. Increasingly frequent droughts in the face of climate change and growing populations along the Rio Grande are certain to put more pressure onto an already exhausted river being dried up on a regular basis. The Department of the Interior estimates that as a result of climate change, the upper Rio Grande will collect less and less water as annual snowpacks shrink and evaporation rates increase. The pressure to stop "wasting" water on a river already being regularly dried up will only continue to mount.

The many and varied participants in the Rio Grande's water war are caught in their own drumbeats and limbo. They are trying to stick it out, hoping that relief will finally come, but increasingly, they're left only with desperate plans such as piping water thousands of miles.

They'll continue using all the water they can for as long as they can, until they can't. It's exactly the same strategy humans have for using all of the planet's natural resources. Because the Rio Grande silvery minnow no longer has a natural ecosystem where it can survive, many biologists now consider it functionally extinct.

LOGGING

My final career destination was Northern California, one of the few areas where the northern spotted owl is making its last stand. One of the reasons this area still has a small population of spotted owls is because tiny remnants of old-growth forests remain in scattered locations on Forest Service land. These older forests with large trees and a closed canopy have abundant prey for the owl, suitable nesting sites, and security from predators and are cooler in the summer and warmer in the winter. Northern spotted owls are almost totally dependent on coniferous forests older than two hundred years.

Old-growth forests that are not a part of a national park or wilderness area are increasingly rare. In addition to the values of ancient woodlands everywhere, these small remaining tracts provide stepping stones allowing birds and animals reliant on old-growth to move across the landscape to the few larger tracts where they might persist. They provide a means of ensuring gene flow between populations and travel corridors to reestablish populations in areas where they may have disappeared.

An 1858 article in the *Humboldt Times* of Eureka stated, "Our immense forests of timber, as a source of wealth, are as valuable as the best gold mines in the State, and they are equally inexhaustible." The rapid rise of California's timber industry was influenced by an abundance of forestlands, the belief that such forests contained an inexhaustible supply of timber, and the proximity of forests to well-populated markets.

In 1852, the first grove of ancient sequoias was discovered in the Sierra Nevada by European Americans. The largest tree in the grove was 1,244 years old, measured 302 feet in height, and 96 feet in circumference. In the typical human response to the discovery of gigantic phenomena, the next year it was cut down.

The demand for sugar pine and redwood quickly spread throughout not only California but other parts of the country. A single old-growth redwood provided enough lumber for a large structure such as a church or hotel. Lumber production in one Northern California county went from 200,000 board feet in 1854 to 40 million board feet in 1874. By the mid-1870s, there were over 300 sawmills in the state, most of which were in Northern California. By 1870, the logging industry had already consumed one-third of the state's forests. By the 1890s, the areas around the Truckee River and Lake Tahoe were so devastated by logging that the mills around the lake were forced to shut down.

Railroads and other advanced technologies resulted in increased timber harvest every decade in the 20th century until the 1990s when President Clinton's Northwest Forest Plan and efforts to save the northern spotted owl slowed harvest levels by more than 50 percent. In 2000, five counties in Northern California still accounted for nearly 55 percent of the total timber harvested in California. Timber harvest has remained fairly constant every year since the Northwest Forest Plan was adopted but is now expected to jump 30 percent as a result of recently passed legislation.

After spending 30 years in the Rocky Mountains and Southwest where trees three feet in diameter were considered monstrous, being in these remnant stands of ancient cathedral-like sugar pine, ponderosa pine, Douglas fir, white fir, and incense cedar provided as close to a religious experience as I've ever had. Their incredible presence is part of a forest continuum spanning millennia, where death feeds life and life feeds death. Few people understand that even when the tree is dead, its life is far from over.

After dying, it remains standing as a snag or stag often for over 50 years. Its center is spongy in places, hollow in others, and has different amounts of sound wood in its periphery. Because of not having needles or branches to catch the wind, it can provide a haven for a number of birds, bats, insects, and other animals in a windstorm. Between their long life as a standing tree and decomposition on the forest floor, they provide food and homes for bats, owls, woodpeckers, and a host of other wildlife. Northern spotted owls also nest in and forage in snags for small cavity-dwelling mammals.

After the snag falls, it provides the nutrients for the forest to continually reboot itself. Whether the trees are dead or alive, the forest is never offline. Both dead and live trees are forever integrated into the forest's continually recharging internet. New trees establish in the organic matter derived from the dead trees where they eventually grow into a towering presence centuries from now. The dead keep the living alive. For millions of years this cycle has repeated itself.

Watching fossil-like black-and-white pileated woodpeckers spend an hour on a downed log feeding on insects reinforces how important dead trees are to inhabitants of the forest. Salamanders, snakes, and insects all rely heavily on downed logs for both food and security. Many species of fungi grow only on dead wood, breaking it down and returning important nutrients to the soil.

In Northern California, complicit with US Forest Service personnel, the timber industry is using their considerable influence with elected representatives to log the small remaining tracts of ancient forests housing northern spotted owls—under the guise of forest health. These small tracts of ancient trees were supposed to be protected from logging as part of the Northwest Forest Plan. The timber industry's success in logging old trees here illustrates how all the trees and forests in the world are in trouble, even the quaintly named set-aside lands.

Considering the long history of logging in California, it's amazing these tracts have somehow managed to evade loggers. When there were a lot of big trees, they cut a lot of big trees. Now that there are only a few big trees left, the timber industry wants to cut them all.

Forest health is a nebulous concept created by humans to mean anything they want. It is usually defined as the production of forest conditions that directly satisfy human needs in the short term. For the timber industry, unhealthy forests means they might die before they can be converted into lumber. Any tree whose end point isn't a board has gone to waste. Most of the public doesn't know anything about forests. Improving forest health sounds like a worthy goal to the uninitiated. Who wouldn't support making our forests healthy?

The logic supporting forest health implies that for millions of years prior to our arrival, forests were always ailing. The only thing that has happened to forests the last 150 years is ceaseless logging and livestock grazing. Now they propose to cure our sick forests with more logging and unabated livestock grazing. Forest health is insanity personified.

Once again, I was asked to pilot my trusty unicorn steed. The unhealthy forests of Northern California were looking for a healer. As self-appointed doctors, the Forest Service and the BLM were going to provide the cure by razing some of the last old-growth forests supporting northern spotted owls. Because of the negative effects of logging to a federally protected species, they needed the blessing of the US Fish and Wildlife Service. Being a little slow to follow their logic, I became a source of extreme frustration for the timber industry and federal agencies. If the trees were humans with the flu, their answer to the problem would have been euthanasia—assuming euthanizing people was a moneymaker.

The timber industry has been quick and effective in capitalizing on people's fear and ignorance of forest ecology and fire behavior to increase logging. They propose managing our forests to eliminate or preclude large wildfires. What they carefully leave out of their information campaign is that managing and logging are the same thing. There is an element of truth to their media blitz. Where there is no forest, there can't be a forest fire. Never once has there been a forest fire in a Walmart parking lot, which is what they essentially strive to create with their logging.

The alternative to complete tree removal is thinning to reduce the amount of trees, called fuel reduction projects. However, under severe fire conditions, wind-blown fire jumps over, around, and through the thinned forest. Some studies have found that thinning can exacerbate fire spread because they open the forest to rapid drying of fuels and allow greater wind penetration.

Eighty-four percent of forest fires are caused by humans, through arson, carelessness, or accidents. There are those who want the public to believe wildfires are caused by unhealthy forests and trees. Their solution is a simple one: eliminate more trees and forests by increasing the timber harvest. Their strategy would be analogous to tearing down banks and stores to prevent robberies and shoplifting. While it would be effective, banks and stores are a part of our lives, so we devise strategies to prevent and minimize robberies and shop lifting. Whether people realize it or not, trees, especially large ones, are the most important organism standing between them and oblivion.

At one meeting I attended with Forest Service personnel, their rationale for a timber sale was the trees might die. It made me feel lucky they were Forest Service employees rather than medical caregivers so those seeking an annual checkup wouldn't end up in the morgue.

Ironically, the Forest Service is in the Department of Agriculture because trees are considered a crop. Everyone in the rest of the agricultural universe knows that after harvesting a crop, the land can become depleted and less fertile. Without replenishing soil nutrients in some fashion, crop productivity dwindles to a trickle. We know this because we've seen it happen over and over again.

No other crop extraction removes more soil nutrients from an area than logging. Old trees are the product of centuries' worth of nutrient uptake from the soil. Only with logging do humans harvest a crop, plant seedlings, and expect a different result than what has been observed with agricultural practices for thousands of years.

The timber industry has a continual barrage of television commercials in geographic areas where logging is prevalent proudly touting their logging as sustainable because they plant seedlings afterward. Sustainable means being able to maintain at a certain level. Impoverished soil conditions coupled with climate change ensure sustainable forests are but one more timber industry ruse. To a public unfamiliar with agriculture or the outdoors, it sounds perfectly reasonable. The timber industry's puffery is a Teflon-coated swindle because there is no accountability for baseless promises extending 75 years into the future, let alone 300.

Recently, the timber industry stole a page out of the water users' handbook of deception. Forests have been playing for millions of years. Fortunately, humans arrived on the scene several hundred years ago and forced them into working. Working forests are the forests that are being logged. In this new, forced role, their natural role as reservoirs of biodiversity and clean air and water are terminated, often for decades and even centuries. They are like birds trying to fly after being stripped of their feathers.

Regarding the role of logging, the extinction of the ivory-billed woodpecker has a striking parallel to the current plight of the northern spotted owl. In the early 1900s, conservationists warned logging was condemning the bird to oblivion. Ivory-billed woodpeckers were once found from South Carolina to Florida and west to Texas in old-growth hardwood forests associated with seasonally flooded rivers and streams, called delta forests. These old-growth forests survive today in small, isolated patches. Unlike the forests in the West, there are no large parks and wilderness areas in the South with more extensive stands of old-growth delta forests. Those are gone forever.

In the early 1930s, a state legislator from northeastern Louisiana shot a male ivory-billed woodpecker in a huge tract of virgin timber along the Tensas River that was owned by the Singer Sewing Machine Company. A researcher from Cornell University verified this bird had been from a small population of ivory-billed woodpeckers in 1937 in what was known as the Singer Tract. That same year, the Singer Company sold logging rights on their property to the Chicago Mill and Lumber Company. During a 1941 visit to the Singer Tract, the researcher observed heavy logging and predicted it would lead to the end of the ivory-billed woodpecker.

In attempts to save the bird, the governors of Louisiana, Tennessee, Mississippi, and Arkansas wrote letters to the Chicago Mill asking them to spare the bird, to no avail. President Franklin Roosevelt directed the Secretary of the Interior to consider how to save the land. All the efforts were rebuffed by both Chicago Mill and the Singer Company.

In January 1944, the last ivory-billed woodpecker was seen at a place on the Singer Tract called John's Bayou. Upon learning John's Bayou may be logged any day, a wildlife artist spent two weeks following and painting the bird. He also saw the logging that cut down the trees used by the last ivory-billed woodpecker. His was the last universally accepted sighting of one of these birds in the United States. When there were a lot of large tracts of old-growth delta forests, humans cut a lot. When there was only one left, they cut it.

Unfortunately, the timber industry's determined efforts to rid the planet of ancient forests is meeting with continual success. They're making inroads every year into eliminating the small pockets of remnant gigantic trees in Northern California and elsewhere.

Painful personal experience has made it clear that no individual can influence the war between ancient forests and humanity. The same results of the conflict have played out repeatedly across the planet for five millennia. It's only now becoming increasingly obvious which side will lose by winning.

INSTANT GRATIFICATION

In the 1960s, a Stanford professor named Walter Michel began conducting a series of psychological studies on hundreds of children, most being around the ages of four and five years old. It revealed what is now believed to be one of the most important characteristics for the success of an individual in health, work, and life.

The experiment began by bringing a child into a private room, sitting them in a chair, and placing a marshmallow on the table in front of them. The child was made an offer: The researcher was going to leave the room, and if the marshmallow had not been eaten upon the researcher's return, the child would get a second marshmallow. If the child decided to eat the first marshmallow before the researcher came back, they would not get a second marshmallow.

The immediate results of the experiment were predictable. Some kids jumped up and ate the marshmallow as soon as the researcher left the room. Others tried to restrain themselves before finally giving in to the temptation of eating the marshmallow. Only a few children managed to wait the entire time and were rewarded with the second marshmallow.

As the years went by, researchers tracked each child's progress in several areas. The children who were willing to delay gratification and waited to receive the second marshmallow ended up having higher SAT scores, lower levels of substance abuse, better social skills, lower rates of obesity, better responses to stress, and generally better scores in a range of other measures. In other words, this series of experiments proved that the ability to delay gratification was critical for success in life.

As a species, *Homo sapiens* is no different than the children who gobbled down the marshmallow as soon as the researchers left the room. Irrigators monetize water, ranchers monetize grasses, and loggers monetize timber, and they want all their money now. The possibility of having more next year isn't given a thought. Politicians have to appeal to their constituents right now. They can't afford to be thinking a decade down the road if they want to be reelected.

Delayed gratification isn't part of our lexicon. We're the species of six-pack abs in just minutes a day. Our lack of long-term success as a species will be, in part, for the same reason that children who couldn't delay immediate gratification were less successful as individuals as they aged.

If you want next century's soil, if you want pure water and clean air, if you want variety and health, if you want stabilizers and benefits we can't even measure, then it requires patience. Forests and grasslands create these things dependably and slowly. Patience and humanity are on different paths heading in the opposite direction.

While the concept of "if there is a lot, take a lot; if there is a little, take it all" followed me around my entire career, I realized it's what humans have done for thousands of years. The scenario played itself out with the passenger pigeon 150 years ago. Even when commercial hunters knew there were few passenger pigeons left, they continued hunting them until they were extinct.

Scientists and paleontologists often repeat the mistake of believing our ancient ancestors didn't know the impact they were having on the mega-faunal species they drove to extinction. While they may have not known the range-wide distribution of many of these species, they knew nearly every resident individual animal within the territory they normally used. When they had the opportunity, they killed every single animal. When it comes to using natural resources, they were no different than we are today.

The concept of "if we don't use it, it will go to waste" isn't limited to grazing, water, and logging but transcends our use of all natural resources. Hunters complain about elk and deer going to waste if they fall prey to wolves and cougars. Fishermen complain about salmon going to waste dying in streams rather than ending up on their plates. Oil and gas deposits are going to waste because they aren't being developed. Ore deposits that aren't mined are going to waste. Empty lots and fields are going to waste because they haven't been subdivided and developed. There are few natural resources on Earth without a large contingent of people wanting to destroy them so they don't go to waste.

The way trappers viewed beaver, ranchers viewed grasses, commercial hunters viewed passenger pigeons, and people in general view water show how we often compete with each other to maximize material self-gratification through endless consumption and production. When it comes to the exploitation of the planet's natural resources, we rarely get beyond our deeply rooted us-versus-them mentality, with *them* usually defined as undeserving others who aren't like us. The attitude of taking it all so others can't have any is pervasive. Invariably, the competition between groups of people comes at the expense of the biodiversity of the planet.

Sometime in the last 20,000 years, humanity traded humility for a heaping dose of pomposity and self-importance. We're here to save the planet from wasting its natural resources. These are the attitudes that will expedite our arrival at humanity's final destination.

CHAPTER 5

SUCCESS

Lights turn on when we flip the switch. Water comes out when we turn on the tap. Groceries magically appear on the shelves of the supermarket. Few understand that everything we do, eat, and purchase creates a ripple of effects, not just the immediate positive we perceive but negative ones across the planet. While this may seem cliché in writing, in action, the ramification for *Homo sapiens* is dire.

HYPOCRISY

Most would consider my way of life extreme environmentalism. My lifestyle includes a small one-bedroom apartment in town, a fuel-efficient vehicle, and utility bills that have hovered around $20–30 a month for over a decade. I have no children, recycle as much as possible, reuse and repair products whenever practicable, minimize my trash, and have carried my own recyclable grocery bags since the early 1990s.

However, my friends are scattered across western North America. And I love experiencing this beautiful world. Whether maintaining friendships or pursuing recreational interests, it requires travel, equipment, and clothing. Smartphones, computers, stereo, and television allow me to communicate, be entertained, and keep up with current events. Numerous surgeries over the years have helped me remain relatively healthy. While my existence is less materialistic than most college-educated people living in either the United States or elsewhere on the planet, it's not sustainable.

The first clue of my own culpability to humanity's problem came when climate change reached the public consciousness. The growing awareness of the devastating impact of the immense quantities of carbon released into the atmosphere each year induced some environmental groups to post "carbon footprint calculators." They allowed individuals to estimate the amount of carbon they generated based on personal lifestyle factors, travel and work habits, dietary choices, and other parameters. Despite my numerous pro-environmental lifestyle choices, I scored about average for carbon footprint as a result of airline flights for both work and personal reasons. That was my first hint regarding the difficulty of functioning in modern society and living sustainably.

Despite these earlier signals, until a decade ago, I still believed mine was a sustainable way of life and if other people mimicked my lifestyle choices, humanity would continue to persist. No matter what a person's lifestyle choices, almost all people think exactly the same way. As long as everyone lives like me, everything will be fine.

I finally realized that the only aspect of my lifestyle that would prevent the catastrophic demise of the ecosystems supporting humanity was not having children. If everyone chose to utilize all the other lifestyle choices I've made, we would still disappear from the planet. Finally coming to grips with my own hypocrisy was a painful realization. Regardless of my background, education, and experience, for most of my life I was blind to the reality of humanity's future and my own contribution to our inevitable demise.

Environmental hypocrisy is rarely black-and-white but almost always shades of gray. Everyone contributes to the carnage on our planet in different ways and amounts. There are plenty of people who sincerely believe they are environmentalists whose damage to the planet far exceeds the average. The opposite is also true for some people who could not care less about the environment actually harming the planet much less than the average person.

The ecological insights gained from a long career in natural resource management helped me clear away the fog shrouding the magnitude of a single person's effect on the planet. For over a decade, I supervised the analyses of federally funded projects to determine their effects on endangered species. A detailed analysis is required by the Endangered Species Act whenever a threatened or endangered species might be killed or harmed.

The analysis requires a biologist to reverse engineer the project so they can understand the various steps of construction or production on imperiled plants and wildlife. Reverse engineering is a way of looking backward into the history of what will transpire. For example, for a timber sale, would loggers have to construct additional roads to log trees? How much road and how wide were the roads to be constructed? What time of year would the logging take place? What type of logging would be utilized? How many and what size trees would be logged? Would there be trees left on-site? If so, how big and how many? How would the logs be moved to the areas where they could be loaded on trucks? How would they be transported from the area that was logged?

This grasp of project details through reverse engineering the project allows the biologist to more accurately estimate the number of endangered plants or animals that might be killed or harmed and ensure that level of harm didn't result in the extinction of the species. These analyses are complex and diverse. They include not only timber sales but livestock grazing, oil and gas development, recreational activities, highway projects, powerlines, pipelines, wind turbines, mining, agriculture, hydroelectric and flood control dam operations, and general construction projects.

The analysis, however, completely ignores impacts to native fish, wildlife, and plants that aren't federally protected under the Endangered Species Act. First out of legal necessity, then habit, biologists compartmentalize impacts and ignore the harm and killing of those plants and animals that aren't protected by the law.

What becomes obvious is there are no projects that don't kill native species, either directly or indirectly. Just because a project doesn't cause the extinction of a species doesn't mean it isn't having negative impacts on biodiversity, sometimes significantly.

After I reviewed many of these analyses, it also eventually became apparent that a person's impact on other species, natural habitats, and the planet is directly proportional to the total natural resources they consume over the course of their lifetime.

Most people are unaware of the processes that maintain their existence. The biological reality is every organism, including humans, live as the result of the death of other organisms, either directly or indirectly.

Nonhuman animals consume what they need to survive. Their consumption is driven by their immediate energy requirements or, at most, their relatively short-term needs, such as when they build fat reserves for winter or to successfully reproduce. Humans consume not only what we need but whatever might satisfy whims and desires within our own personal financial constraints, and sometimes not even within those.

The enigma within the conundrum is that success in human society is rewarded by allowing the individual to increase their consumption of natural resources. Bigger houses and families, larger and better vehicles, frequent travel, sizable wardrobes, superior foods, and extravagant hobbies all lead to the consumption of more natural resources. The only way for a person to minimize not only the number of organisms they kill but their effect on our deteriorating planet is to consume fewer natural resources.

For everything we do, there are native plants and animals that have died or are dying to make that a reality. This single premise will ensure humanity can neither prevent the collapse of Earth's ecosystem nor prevent our acceleration toward an inevitable demise. Almost everyone wants to be successful and rewarded for their success. Who among us wants to be poor and unsuccessful?

The average home in the United States is now 74 percent larger than it was in 1910. The average personal living space increased 211 percent during that same period. An average home today requires 22 mature pine trees to build compared to the 13 necessary a century ago. An average mansion needs 67 mature pine trees to construct. The mansions of elite athletes, movie stars, CEOs, and other rich people require cutting

127 mature pine trees. This figure only pertains to the framing of a house. Features like hardwood floors, cabinets, and staircases can easily double the number of trees cut down to make that home a reality.

The typical American uses 41 percent of their energy on space heating and 35 percent on appliances, electronics, and lighting. An average US residential customer uses approximately 909 kilowatt hours per month of energy. Large mansions can use approximately 31,000 kilowatt hours per month. Even from this simple illustration, it becomes obvious the more "successful" a person is in society, the greater their contribution to defiling the planet.

It should come as no surprise that wildlife has declined by 60 percent since 1960. During the same time, the human population has increased by 150 percent. It is a testament to the perseverance and tenacity of the nonhuman organisms still surviving, given the current level of human impacts to both them and the places they inhabit.

Every human needs clothing, food, water, and shelter. Nearly everyone wants a vehicle, television, stereo, computer, smartphone, swimming pool, mansion, closet full of clothes, pantry full of food, fancy appliances, extra water for nice long showers. The very successful people generally get everything they want. Less successful people have smaller houses, their vehicles aren't as expensive, they forgo swimming pools, and have fewer clothes. Poor people scrape by in small houses or apartments, with few clothes and barely enough food to survive. Because they use fewer natural resources, the impoverished—particularly those with few or no children—are the true environmentalists in modern society. In fact, it has nothing to do with them being environmentally conscious. They would almost certainly prefer to be rich and "successful."

Our political leaders have bent over backward to fight the perception that wealth can be a curse, especially when it becomes an end in itself. Wealth is to be celebrated and admired, however nefariously it might have been obtained. Anyone who questions the right of the wealthy to enjoy the fruits of their labors or their good fortune is indulging in "the politics of envy." Even poor people often believe the rich richly deserve every perk they enjoy.

To better understand why "success" and environmental destruction are synonymous requires doing the same thing we did with the earlier example of a timber sale, namely, reverse engineering a good or service. The timber sale only covered harvesting the wood and transporting it off the area where it was cut. There are almost always many additional steps between extracting and transporting natural resources and having them available for consumption by the public as a good or product.

While constructing and heating homes may be more obvious, even the seemingly benign activities we undertake have negative effects on the planet. The processes involved in extracting, producing, transporting, and selling the goods and services we enjoy have become increasingly complex. A basic comprehension of these processes is necessary to understand how we all contribute, usually unknowingly, to undermining ecosystems that we and all of the planet's organisms depend upon.

RICE

As an example, when we sit down to a simple meal of a bowl of rice, we're sustaining ourselves with nourishing food. It is not our intent to destroy natural ecosystems or to contribute to climate change and deforestation. However, that is exactly what has happened. The environmental impacts of growing, harvesting, processing, storing, transporting, and distributing the rice in your bowl or any other cultivated crop extend across the planet with irrevocable consequences.

Rice is a cultivated crop. To provide the necessary conditions for its successful cultivation, natural landscapes have been converted to rice fields. *Conversion* is a vanilla word meaning clearing forests, draining wetlands, destroying natural grasslands, and reducing biodiversity. Worldwide, an area the size of India is devoted to the cultivation of rice, some 400 million acres.

During the initial conversion process, wildlife previously using those areas are either killed outright or displaced. After an agricultural crop is established, native wildlife and wayward plants living in adjacent areas attempting to use or encroach on the crops are driven out. If they persist, they are exterminated in perpetuity by the farmers.

Grasslands, forests, and marshes harbor an incredible variety of plants and animals that provide natural checks and balances, holding individual species in check. An insect that favors or depends on a single species of plant in the forest, wetland, or grassland can't build up large populations if that plant species is only a small part of the natural community. Cultivated fields are by design devoted to a single species. Extensive areas planted to a single crop sets the stage for explosive increases in specific insect populations and invasive weeds. This paves the way for the need for insecticides and herbicides.

Rice is cultivated in flooded fields or paddies. Moving water from the river to fields requires a massive network of dams, canals, and other works. Dams and irrigation projects have far-reaching environmental impacts. Their construction permanently removes native wildlife habitat and permanently alters the landscape.

River systems are the areas of Earth's highest biological diversity and bioabundance. They also have the most intense human activity. Freshwater biodiversity is in a state of crisis, the consequence of centuries of humans exploiting rivers with dams, water diversions, and pollution. Agricultural diversions across the planet have caused the extinction of many plants and animals and jeopardize the continued existence of numerous others. In addition to the waterway itself, the river's course hosts a ribbon of vegetation along its banks, nurtured and supported by the river's flow.

When waters are diverted for irrigation, there is no longer adequate moisture and sediment necessary to sustain the numbers and kinds of plants and animals present downstream of the diversion. During the irrigation season, reduced river flows can no longer support the survival of many of the native fish, wildlife, insects, amphibians, reptiles, and native plants.

Rather than a complex ecosystem supporting hundreds of species, the area influenced by water may support only a fraction of the original abundance and diversity. Reducing the amount of water or the timing of when water is available can prevent fish from successfully reproducing. Newly hatched fingerlings are unable to survive because many of the areas they need in the early stages of life are either unsuitable or unavailable.

Irrigation projects move water from rivers in concrete-lined canals and ditches. Canals and ditches and cemented riverbanks are stark environments hostile to native populations.

When rice is planted, insecticides, fungicides, and herbicides are routinely used to maximize production. Chemically laden drain waters from irrigated fields contaminate soil and eventually work their way back to the river or into the groundwater aquifer, where impacts accumulate and reach far downstream.

Rice is an intensive crop and requires heavy applications of fertilizers. Phosphate- and nitrate-laden drain waters from rice fields disrupt natural environmental balances and may encourage blooms of algae and other microorganisms, choking the oxygen from waterways and killing fish and other animals.

Rice grows mostly in flooded fields called paddies. The water blocks oxygen from penetrating the soil, creating ideal conditions for bacteria that emit methane. The extensive use of water in rice farming increases methane emissions globally by 14 percent. Methane is 28 to 86 times more potent as a greenhouse gas than carbon dioxide and is responsible for 10 percent of all emissions contributing to climate change. Harvesting machines, transport vehicles, fuel, roads, and highway infrastructure are necessary for cutting, stacking, handling, threshing, cleaning, and hauling the rice. The rice goes through a milling process before being packaged and shipped to ports for export or to distribution facilities.

Roads and highways are used to transport harvested rice to processing plants and ultimately to markets or shipping destinations. Roads and highways have impacts on wildlife and their habitat that are dispropor-tionate to the area of land they occupy.

Roads are increasingly prevalent in today's world as human development expands and people increasingly rely on vehicles for daily transportation of food, products, and themselves. This has altered the landscape impacting the planet and wildlife both directly and indirectly.

As of 2018, there were 40 million miles of roadways on the planet. This mass of highways and roads would completely cover the country of Zimbabwe. The dark pavement absorbs heat and sunlight, increasing surface temperatures by as much as 20 degrees Fahrenheit. The heat dries out adjacent vegetation creating combustible fuel when cigarettes are thrown out of automobiles or power lines paralleling highways are downed by wind or fallen trees. These factors have given rise to many large wildfires.

In 1960, there were roughly 62 million vehicles in the United States. By 2018, there were 272 million vehicles. Between 1960 and 2018, the US population nearly doubled, but the number of vehicles more than quadrupled. Instead of the one family car in 1960, everyone in the family has a car in 2018.

Large freeways and improved highways rarely start out that way. They almost always begin as a small dirt two-track road. As more people and vehicles use the road, they are gradually improved. Each time they are paved or widened, there is more construction, more organisms are killed, and more habitat lost. Fewer and fewer animals can safely move across the highway until at some point, the traffic is so heavy or the road so wide it becomes an impermeable barrier or certain death trap.

Vehicles hauling the rice kill wildlife crossing roads and highways. Most people have become so calloused to dead wildlife routinely killed on highways it hardly even registers in their consciousness. Big-game hunters routinely complain and mewl about the effects of wolves and cougars on big-game populations while driving by dead deer and elk slaughtered on the highways without giving it a second thought.

Roads and highways are ribbons of small-mammal and bird carnage. An estimated 1 million vertebrates die on roads every day in the United States alone. Vehicle collisions are the leading mortality source of many wildlife populations. It is a significant component in the risk of extinction in some species. For animals already suffering from the effects of climate change or urban expansion into their habitat, the additional mortality associated with roads and highways can ensure they never recover.

Roads divide large and contiguous areas into smaller, isolated patches. At the point when traffic is so bad that animals are either killed trying to cross or avoid the roads altogether, those that survive can suffer by not being able to access areas necessary for their survival. For example, in times of drought, roads can prohibit animals from reaching water. Roads and highways can disrupt migration patterns, leaving animals striving to survive in unsuitable areas during certain times of the year.

Pollution into the environment from roads and highways is never-ending. Debris from tires, deicing salts, trash thrown out of vehicles, and oils all have the potential to harm fish, wildlife, and water quality.

So far, our reverse engineering example of a bowl of rice has only provided a cursory overview of impacts from water diversions for irrigation and from the transportation infrastructure to bring rice to markets. We haven't addressed the impacts of mining to obtain minerals to make materials for fabricating the rice-harvesting equipment or the manufacturing process for making and assembling the vehicles and equipment. There are also the plastics and packaging industry impacts, oil and gas development for fossil fuels, and chemical plants for the production of the insecticides, herbicides, and fungicides. A detailed reverse engineering analysis of all aspects of a bowl of rice would require a book so lengthy it would make *War and Peace* look like a pamphlet.

Virtually no one goes into a grocery or department store and marvels at all the native plants and wildlife killed to ensure access to a plethora of items making their lifestyle comfortable. At the same time, not only is there no desire to slow down or stop our assault on the planet but the opposite is also true. Almost everyone wants more. A healthy economy requires the increased spending and consumption of goods and services by consumers, more natural resource extraction, and ever-increasing amounts of trash and bioaccumulating toxins.

While one bowl of rice may not seem like a big deal, enough rice to meet the demand of nearly 8 billion people on the planet is momentous. For the planet and native species, it is death by a thousand cuts or, in this one example, billions of bowls of rice.

Whether a bowl of rice, a cup of coffee, flushing the toilet, going to a movie, or buying an article of clothing, there is nothing we do that if reverse engineered wouldn't reveal negative effects to native plants and animals and another contribution to our deteriorating planet. There is almost nothing we do, consume, or purchase that when reverse engineered doesn't require at a minimum deforestation, mining, desertification, and oil and gas development. It also means that everything we do contributes to climate change, toxins in the environment, and the loss of biodiversity on the planet.

There is another common and equally erroneous way humanity rationalizes not harming wildlife. When constructing new roads, highways, cultivated fields, homes, subdivisions, shopping malls, or any other structure that permanently removes natural habitat, people believe the animals residing there can just move.

Most people don't appreciate, and therefore they ignore, placing any value on all the animals, plants, and organisms that can't move homes. Organisms like small mammals, native plants, earthworms, soil bacteria, fungi, and insects are easily objectified, aren't worthy of mention, and are killed outright. How many animal rights groups spring to the defense of mice, spiders, shrews, and beetles?

However, most human beings also don't understand the effects to those organisms mobile enough to avoid the direct effects of a new construction project. Imagine living in a confined, self-sufficient community with 100 homes. It has adequate housing, food, and water for all the people to live there comfortably, but no more. A natural disaster permanently destroys 50 homes along with all of their food and water, but all the people living in them manage to survive.

All these people are forced to move somewhere else within the self-sufficient community. Imagine the reaction of those people who weren't affected to suddenly being asked to give up 50 percent of all their food and water when they had barely enough to survive prior to the disaster.

The people who were directly affected by the tragedy and those who weren't are initially crowded, starving, and thirsty in an attempt to accommodate everyone. Whether people die of malnutrition, thirst, or territorial disputes, it is unlikely to end well for many of the people in the community. In fact, after a time, it is entirely likely the self-sufficient community will reach a new equilibrium supporting the same number of people as lived in the 50 unaffected homes prior to the disaster. Half of the population will be lost in spite of no one dying in the initial disaster.

It is no different for displaced wildlife mobile enough to move into adjacent areas to avoid the direct effects of construction activities or converting native ecosystems to pastures and croplands. After the initial crowding and territorial disputes, a new equilibrium will be reached that, at best, only supports the number of animals that were in the adjacent areas prior to the construction.

The populations of animals that inhabited the construction area, croplands, or pasture have been lost forever. This is the primary reason why habitat loss rates as the number one reason for the precipitous decline in wildlife populations and biodiversity in the last 50 years.

The intricacies of the processes make explaining reverse engineering difficult. Describing the effects of each of the processes on the environment can be even more onerous. It is this complexity and lack of understanding we've created in almost all aspects of our lives that ensures few clearly perceive their role in contributing to the deterioration of our planet.

DISCONNECTION

Today, most people in industrial societies don't need to know much about the natural world in order to survive. You need to know a lot about your own tiny field of expertise, but for the vast majority of life's necessities, you rely blindly on the help of other experts, whose own knowledge is also limited to a tiny field of expertise. Our limited worldview along with the complexity of all the various processes allow us to create distance between us and the plants and animals we routinely destroy and, by proxy, the planet we affect. There is almost nothing people eat, drink, purchase, or do that doesn't cause an ecological disruption. There is nowhere we travel and there are very few interests we pursue that doesn't negatively impact the environment.

Research has shown that physical and psychological distance between two people, one who is in pain, is a decisive factor for the strength of an empathic reaction in the other. We see these phenomena playing out in our communities and world every day. When a mass shooting or a tornado directly affects a community, the people left alive in the city or town where it occured are affected much more profoundly than those who live in other places across the country.

Similarly, someone who watches birds nesting in a tree near their home over a cup of coffee every morning will likely be angry, sad, or frustrated if that tree is cut down. However, if the same person goes to Home Depot to buy lumber for a new cabinet, it doesn't even register that wildlife died during the logging, transport, and milling of the materials for their home improvement project.

The amount of paper towels utilized by members of gyms and health clubs to clean sweat and germs off cardio and weightlifting machines requires an incredible amount of logging to satisfy. There are few, if any, gym members who make the connection they are killing wildlife and contributing to deforestation and climate change in their pursuit of a germ-free workout. Either perceived or actual physical and psychological distance makes us callous to the effects of our lifestyle decisions on the planet.

As a result of expanding urbanization, humanity is becoming increasingly disconnected from the habitat that provides our sustenance. More and larger urban areas are a part of our lives. Most people don't have the opportunity to either experience or understand what is happening to our habitat. Being completely disconnected from all aspects of the natural resource extraction process that facilitates survival doesn't mean it isn't happening. Instead, the disconnection creates the physical and psychological distance necessary for people either not to care or not to recognize our role in a swiftly deteriorating planet.

Growing families, businesses, or the economy all requires greater natural resource extraction and consumption. Waste and growth are inseparable. Vehicles, electronics, appliances, furniture, water, paper products, plastics, chemicals, and clothing are but a few of the items routinely purchased and either used up or thrown away. Without waste, our economy would collapse and growth would stall. Imagine what would happen to electronic companies if only one television or one smartphone was necessary in a person's lifetime.

Most companies and governments encourage waste in a variety of ways, so people are required to buy more of their products. People and companies extract natural resources to facilitate economic growth by grazing cattle, logging forests, mining, diverting water from rivers, developing oil and gas, and a host of other activities. While the waste generated from these activities pollutes the land, air, and water, it is fostered and encouraged to inspire additional economic growth and prosperity.

The planet's human population is currently growing by a country the size of Germany, annually. The strain on our planet from the current global population is already manifesting itself in profound ways, including climate change, deforestation, desertification, infectious disease, geopolitical instability, mountains of trash and toxic substances, and loss of biodiversity.

Even if there were zero human population growth beginning today, all those threats would worsen. This may seem counterintuitive but can be illustrated with a financial analogy. To have a child, a couple decides they need to borrow $1,000 at 15 percent interest. At the end of a year, they now owe $1,150. They love being parents so much they decide to have another and borrow another $1,000. They do this each year for five years. At the end of the five-year period, they decide their family is big enough. No longer borrowing money doesn't mean their financial problems evaporated, and in fact, now they are dire. They owe $7,700 dollars on the principle and $1,150 dollars annually just on the interest. Their financial situation will gradually worsen because after five years of borrowing money, they are unable to even pay the interest.

Similarly, as our population has grown, humans have been liquidating the planet's natural resources for thousands of years. The effects of our current population on forests, grasslands, biodiversity, clean air, and water are catastrophic. Zero population growth means the planet would still be trying to support the current population, over 8 billion people. Humanity can no longer pay even the interest. We've already spent much of the planet's clean air, water, and biodiversity with no mechanism for repayment.

Zero population growth isn't going to happen. We can't curb our evolutionarily programmed need to grow and reproduce any more than we can stop the sun from rising. Like all populations of organisms on the planet, the number of humans will continue to increase until it's no longer possible. When that day happens, Earth will be a truly inhospitable place.

Reverse engineering all of the camping equipment, clothes, gear, vehicles, and boats utilized by people who hunt and fish would show hunters and fishermen kill more wildlife getting ready to go hunting and fishing than actually doing it. Similarly, when reverse engineering the sport of golf, the average golfer kills more animals in total than the average hunter or fisherman. Not only is there a plethora of equipment but golf courses permanently remove native ecosystems and require an incredible amount of water to maintain, contributing to deforestation, desertification, loss of biodiversity, and loss of available clean water.

Success is now its own industry. There are newsletters, podcasts, magazines, self-help books, consultants, and motivational speakers whose entire existence is based on helping others be successful. Many equate success with their own personal financial value.

Many people seek more wealth not out of a perceived material need but out of a real psychological need. It's a kind of mental instability that overtakes a person, leading them erroneously to believe their human worth can be measured in money, and then to torture themselves wondering why there is never enough to do the job. The most common reaction of the human mind to achievement is not satisfaction but craving for more.

We're convinced that to be happy, we must consume as many products and services as possible. Our life won't be nearly as fulfilling with an iPhone 12 as it would be with an iPhone 15 Pro. Success is always just out of reach. Every television commercial assures us that consuming some product or service will make our lives better.

Most nations define success in the same way. Economic growth is the market value of goods and services produced by an economy over time. Those nations with a healthy, growing economy have an ever-increasing amount of construction and manufacturing, which inspires more deforestation, desertification, garbage, chemicals and toxins, and oil and gas development. Nations and individuals seek validation from their counterparts through increased materialism. Perceived success breeds self-deception and makes it possible to ignore the looming crisis. The self-preservation of our species has become roadkill on the highway to success.

CHAPTER 6

ACCELERATION

The human population and their domesticated plants and animals are rapidly increasing in number. This, coupled with escalating consumption by individuals and countries, is the perfect recipe for accelerating to environmental catastrophe. Ecosystems, species, and even individuals that have existed for millennia are quickly disappearing.

ASPEN

Pando, a 43-hectare aspen grove in Utah, may be the most massive living organism on Earth. The grove consists of 40,000 genetically identical aspen trees connected by a single 14,000-year-old root system. Researchers have named the grove Pando, Latin for "I spread."

Pando was witness to mammoths, mastodons, ground sloths, and saber-toothed cats. It has survived drought, fire, extreme heat and cold, and the ever-present depredations of insects and disease. And like many quaking aspen groves throughout western North America, Pando is dying. Aspen groves have been disappearing for the last century. The main culprit is domestic livestock grazing, although deer and elk are also contributors to their demise.

The leaves of aspen trees hang from slender, flattened stalks, called petioles, that catch even the slightest air movements. This feature ensures the leaves tremble even in a gentle breeze, leading to their common name, quaking aspen. A walk through an aspen grove under a dappled canopy of shimmering leaves is restorative to all but the most jaded souls. Autumn displays of gently trembling gold and yellow leaves are western North America's counterpart to New England's famous fall colors. They inhabit some of the most spectacular landscapes on the planet. Their light-colored bark with black horizontal scars is distinctive among trees in western North America.

The unique adaptation of many trees derived from a single root system allows them to spread out into drier areas normally receiving more sunlight and at the same time tap into wetter, more shaded areas to quench their thirst. The aspen in poor soils can receive nutrients from those in rich soils. Millions of years before humans migrated to North America, aspen groves were practicing socialism—wealth redistribution infused with a robust social safety net.

Unlike the larger aspen groves in Colorado and Utah, those in Idaho were more modest in size. Their small size made them an even greater magnet for cows, sheep, elk, and deer. Every aspen grove was a five-star restaurant for ungulates, both wild and domestic. The palatability of aspen to wild ungulates almost certainly extended to megafaunal browsers and grazers. Because of their greater size, mature trees and entire clones were likely utilized by mammoths, giant ground sloths, and other megafaunal species. Many aspen groves had already disappeared before I began working there in 1980. Many of those remaining were in exactly the same condition that Pando is in now. Without serious and aggressive intervention, the small remaining aspen groves will die and be lost forever.

One particularly beautiful grove in the Pahsimeroi Valley of central Idaho of about three or four acres in size was completely isolated. The nearest aspen grove in any direction was several miles away. Those aspen patches that had once existed in closer proximity had long since perished. The grove's location was in a remote area accessed with difficulty by a four-wheel-drive vehicle. It was barely clinging to existence where its counterparts had already succumbed to the ravages of livestock grazing.

Over a three-year period, about five miles west of this aspen grove, I excluded livestock from a stream and the wet areas adjacent to it with fencing. The stream contained bull trout, a now federally protected fish. The lush vegetation that rebounded after excluding livestock was regularly visited by elk. They could easily access the lush, green area by jumping over the fence where cattle could not. Herds of up to 40 elk were regularly seen inside the fenced areas, as well as an abundance of other wildlife attracted to the only area in the entire valley along a stream not obliterated by cattle.

Whenever I ran into the rancher who had the federal permit to graze the area, he would become angry and chastise me for frittering away taxpayer money by keeping his livestock from using all that feed now going to waste on wildlife. His ire was contagious, with friends and colleagues grazing livestock in adjacent federal allotments in the valley sympathetic to his cause.

Regardless of heavy elk grazing, both the quantity and quality of vegetation continued to improve inside the fenced area. This continued vegetative improvement culminated in the ranchers and their advocates losing their scapegoat for deteriorating conditions along streams in the region, namely elk. No longer able to pass the red-face test when blaming elk for overgrazing along streams in central Idaho had them livid. The improved vegetation inside the fenced area made it obvious that cows were the problem.

Their collective fury at the person responsible for this tragedy blossomed, and soon the entire state's livestock industry had labeled me as persona non grata. This was all transpiring contemporaneously with my attempt to save the aspen grove.

Aspen stands resprout after being killed by fire or chewed down by beaver. However, it is also possible to cut mature trees with a chainsaw, particularly in winter when more of their energy reserves are stored in the root system, and still achieve the same desired effect of encouraging regeneration of the stand. Unlike modern "forest health" treatments, all trees and nutrients are left on-site so no human wealth is extracted under disingenuous and nefarious rationalizations. My problem with saving the clone was twofold. The first was accessing a high-elevation remote area in the winter to cut the overmature, dying trees. Even more difficult was keeping not only livestock away from the seedlings but deer and elk for at least four or five years until the seedlings had grown into young trees and were no longer susceptible to being killed by a smorgasbord of ungulates that find young aspen an epicurean delight.

For three weeks during the winter, I chained up a four-wheel drive and drove through 12 to 24 inches of snow in the remote area to fell all the aspens. This step of the project went without a hitch. The weather was as cooperative as winters in the northern US Rockies can be.

The second phase of the project was much trickier and more complex. I had learned from painful experience that to withstand incredible livestock pressure, the fences protecting lush vegetation needed to be more fortress-like than fence-like. The cows would congregate en masse at the fence and gaze lovingly at the succulent vegetation inside. Immediately outside the fence was complete annihilation of every green organism the cattle could lap a lip around. The cows would make a human army envious of their ability to storm these fences at all costs.

The fence-fortress was designed accordingly, and I had the fence materials delivered to the site in late spring six weeks before livestock were scheduled to arrive in the area. It was going to take about three weeks of hard work to construct the fort necessary to repel the ungulate battalion.

There is almost no fence design elk and deer can't navigate. To stymie their efforts to eat the young saplings, I felled the aspen so they crisscrossed each other, creating small areas that even deer and elk couldn't access. My hope was that by the time the fallen aspen rotted away and the livestock and stockmen siege had destroyed the fence, the new aspen clone would be 10 to 12 feet tall and viable for another 40 to 80 years.

A week after the fencing material was delivered, I arrived with my crew to begin constructing the fence and found that someone had burned all the fence materials. The arsonist was never apprehended, but there was a host of ranchers in the valley with the means, motive, opportunity, and dream to thwart anything I attempted. Fortunately, through some luck and fiscal gymnastics, I was able to hurriedly reacquire fencing materials and get the fence constructed before the livestock onslaught began. Saving this aspen stand from a certain oblivion was one of my few unqualified career successes. In 2020, there was an island of 20-to-25-foot-tall healthy aspen within the now compromised fenced area.

TROUBLE

These semi-successful indiscretions of excluding livestock from tiny areas on the landscape eventually led to me being reassigned to a position in bureaucratic purgatory in the southern part of Idaho. My newly assigned hades was a fisheries biologist position in an area with few fish. I was a doctor looking for a patient. A position without many demands left me plenty of time to pursue other endeavors and get into trouble in new and unforeseen ways. One was kicked off by that most dreadful of things, the land-use plan.

The Bureau of Land Management (BLM) was required by law to develop land-use plans. These plans are supposedly a compromise between all the various competing resources and involve the public. They were and probably still are a big deal within the agency. Huge amounts of time and effort were spent writing land-use plans. Career advancement in the agency was often premised on staying in the office for years writing them. As soon as the plans were completed, they were placed on shelves to gather dust in obscure storage areas, and any compromises in the plan for fish and wildlife were quickly forgotten. Staying in a nice, comfortable office writing plans, avoiding conflict, and padding resumes was a torture I avoided at all costs.

But I was, of course, to abide by what was written in others' plans. One recently completed plan expressed concerns for a rare fish, which resulted in old-growth ponderosa pine in one small drainage being excluded from any future timber sales. That was the only concession for wildlife in an area encompassing hundreds of thousands of acres of public land. In the agency spirit of ignoring concessions for wildlife and the environment, the BLM manager for the area decided he wanted it logged anyway. He directed staff to conduct a field site visit of the area to begin an assessment of environmental effects of the timber sale.

Five of us, representing different disciplines, spent a day in the area and unanimously concluded logging was a bad idea for a host of reasons. We jointly wrote up a report with our conclusions, and it was given to the manager. Though it matched the land-use plan, it didn't match the predetermined managerial decision, and it wasn't well received.

Over the next week, all the employees who went on the site visit were bullied and intimidated, and their careers were threatened. One employee came out of a meeting with the manager in tears. The result of this effort was everyone from the site visit changed their analysis so that the timber sale could move forward, with one notable exception—me. The managers didn't realize more accomplished bullies than they had failed to sway me in the past and their determined efforts were wasted. It also became increasingly difficult to discipline someone who was already being punished.

I wrote up my analysis, and at the same time carefully documented their intimidating actions and overt disregard for their own regulations. Because the managers didn't want anything negative said about the timber sale in the environmental analysis, they rewrote my analysis to make it appear the effects were benign and ascribed the rewrite as mine. In this way, the document would be seen as credible by the public because the analysis was blessed by the fisheries biologist.

This all transpired the year Bill Clinton became president. At the time, he had just selected a new BLM director who I thought would be chagrined by the actions of the local managers. I created a packet with all the information about the timber sale, the agency's actions, and my unedited analysis and explained what had transpired. The day after I snail-mailed this information to the new director, he was forced to step down because of some indiscretion in his past, and the old guard was immediately back in charge. They weren't amused when my envelope arrived.

Over the next month, rumors flew as to what new punishment lay in store for me. Being fired or reassigned to one of the least desirable offices in the agency were the options that received the most consideration. Although I was already a pariah, my cubicle became positively toxic, with everyone avoiding me. Serendipitously, the state wildlife biologist secretly admired what I had accomplished over the previous decade. He finally convinced the state director to transfer me to a different office in a new program to work with the state fish and game department and the US Forest Service to conserve rare species.

In my 33-year career, this was my one big lucky break. After being whip-sawed for nearly a month about possibly being fired or transferred to an unpleasant destination, my second punishment landed me in the best position of my career. After spending a suitable period in my carefully orchestrated grave in southern Idaho, like Lazarus I had reemerged to become a somewhat wiser irritant to a bureaucracy intent on destroying the planet and accelerating humanity's path to extinction.

PIKA

A full decade prior to this near exodus from the government, I had back-packed into some high alpine lakes in the Idaho Rocky Mountains for a weekend of fly-fishing. It was a cold, crisp, clear evening, and I was setting up my tent in the last vestiges of trees clinging to the minuscule pockets of soil at timberline. Suddenly, I heard a loud bleat. My immediate thought was there was either an injured bighorn sheep lamb or a mountain goat kid nearby.

My tent pitching was quickly put on hold, and I moved in the direction of the sound. Despite carefully scanning the area where the sound seemed to have originated, I couldn't see anything. Erosion from the bare granitic cliffs and steep slopes above me had deposited rocky scree at their feet.

There were little patches of sedges, grasses, and flowers at the base of the mass of small- and medium-sized stones. After standing silently for about a minute, I heard the bleat again and saw a quick movement in the scree.

I sat down out of sight and was soon rewarded with my first look at an American pika, known in the 19th century as the "little chief hare." After a short time, several others became visible moving in and out of spaces between the rocks and boulders. Their calling sounds were disproportionate to their size. It reminded me of talking with someone on the phone who has a loud, deep voice you think is attached to the body of a massive professional football linebacker. When you finally meet the person, he turns out to be a small, slight guy who can barely coax a scale to 120 pounds, if his pockets were full of rocks.

Here was this small mammal living at the very limits of what elevation allows in some of the most beautiful country on the planet. I've seen them in a number of places since that evening but have always felt pikas are the busiest and most endearing denizens of the highest elevations in North America.

Despite their high-elevation mountain home, pikas don't hibernate; therefore, they have greater energy demands than other animals living in mountainous areas. They eat herbaceous vegetation when it is obtainable or cache food in hay piles to use for a source of subsistence in the winter. Their ability to stockpile sustenance in hay piles is limited to the summertime when vegetation is available and growing. When stockpiling plants for the winter, they will often make 100 trips per day. Pikas harvest different varieties of plants in a deliberate sequence and seem to assess their nutritional value. Forbs and tall grasses tend to be cached more than eaten directly.

Pikas have a small, round body about six to eight inches in length. They are cousins of rabbits, but unlike their distant relatives, they have short legs and round ears. Their color varies from gray to cinnamon-brown. Pikas utilize both calls and songs to communicate among themselves. A call is used to recognize individuals or warn when a predator is lurking nearby, and a song is used during the breeding season and in the autumn.

By 2006, climate change was already adversely affecting several rare species in the Southwest. In the course of analyzing effects to those critters, some research showing dwindling pika populations somehow found its way to my desk. Because pikas can die in six hours when exposed to temperatures above 78 degrees Fahrenheit, the research showed they were already disappearing at the southern portion of their range where they were most vulnerable to climate change.

It was immediately obvious to me; pikas would be one of the first casualties of global warming. They lived at the highest elevations where vegetation was still available, so they couldn't move any farther upward to escape the ever-increasing temperatures. Because of their small size, they would be unable to move farther north to outrun the warming climate. More than a few nights, I lay awake at night saddened, trying to figure out a way to prevent the extinction of one of the many animals I had befriended over the years.

Gradually, I began formulating a strategy that might give them a slim chance of persisting for at least a few more decades. It would involve capturing them and hopscotching them farther north. The shrinking glaciers that were also the victim of climate change could in their demise provide new pika habitat where they had receded. Since no mammals had previously lived there, it would minimize any risk of disease transmission from translocated pikas to other small mammals in the new area. My strategy would require interstate and even international cooperation, but it seemed the only possibility for their continued survival.

The complex proposal to rescue pikas was cobbled together on the sly, mostly on my own time. My agency would have been chagrined to learn effort was being taken to conserve a species they didn't care about. As a result, it took until 2010 for it to be crafted with the aid of a biologist who was working for me at the time. By then I had transferred to Northern California.

I knew the only chance of the proposal being considered involved skipping over the multiple layers of bureaucracy in the agency, or its fate was a certain and swift death on either a forgotten shelf or in the old circular file.

At the time, the agency had a science advisor who was one level below the director. I decided he would be my one and only chance of success. With all my hopes of providing pikas with a temporary stay of execution, the proposal was emailed to him directly. A week later I received a polite response thanking me for the proposal and copied to an underling. His response was the equivalent of thanking me for sending him a turd and warning me to prepare myself for the fallout.

Predictably, three or four days later, my supervisor stormed into my office raging about not going through proper channels. I have no doubt his supervisor did the same to him ad nauseam right on up the chain of command. Nowhere in the numerous layers of bureaucracy was anyone perturbed by the precarious plight of pikas or did anyone even ask to see the proposal that caused the firestorm. While the reaction is typical of large bureaucracies everywhere, this is America's premier fish and wildlife agency, the US Fish and Wildlife Service.

Through painful prior experiences, I knew long before sending the proposal that the outcome would likely be exactly what transpired. Because of an overwhelming desire to assist the animal, I had to take the risk of failure. My efforts on behalf of pikas along with dragging my feet on the approval of logging ancient forests resulted in me being sentenced one last time to bureaucratic perdition. My last assignment prior to retiring was filing a report on the water quality of our office complex's drinking water. *Game of Thrones* must have been inspired by someone working in a natural resource agency trying to slow the rapidly accelerating vehicle to our extinction.

Pikas are just another species that will vanish as Earth's current "custodians" watch, worrying about chain of command, going through proper channels, and economic growth. Like almost all species, pikas are expendable because they can't be monetized.

LEADERS ON THE PATH TO NOWHERE

Many erroneously believe federal agencies in America like the US Forest Service and the BLM are slowing or even stopping the vehicle transporting humanity to extinction. In fact, their false facade turbocharges our race car. Most natural resource employees sincerely believe ecosystems are resilient, not fragile; all ecological problems can be fixed with money or technology; a strong economy is commensurate with a healthy environment; and business as usual is the best course of action. The agencies are populated by sycophants to the industries they are tasked with regulating. Most countries have no false front and don't even pretend concern for the planet. The result is the same.

The conflicts I encountered during my career illustrate how no amount of hard work, personal risk, or willingness to be constantly exposed to criticism and conflict can slow the rate the planet is deteriorating. As a child, I was taught that *please* is the magic word. In natural resource management, the magic word is *yes*. Yes, you can mine, graze, log, drill, irrigate, divert, develop, recreate, blade, and build. It was the only word necessary to keep the majority of humanity joyful both within and outside the agency. Fish, wildlife, trees, other plants, rivers, springs, lakes, and streams never once came into the office screaming at a natural resource manager. There has never been an instance where they gave an employee a bad annual review.

There was the occasional annoying environmental group. Stonewalling enviros, who were without the vast resources of the federal government at their disposal for a cause they considered golden, was mere child's play and a badge of honor for managers long divorced from ethical considerations. A blizzard of paperwork, endless coordination meetings, or litigation were simple solutions in draining the limited resources of environmental groups.

Employees either had long ago lost their resource ethic or had been intimidated into silence. This resulted in an agency atmosphere where the only risk was from shoulder injuries from patting themselves on the back. After all, everyone agreed they were doing God's work. There were extended periods of time when it could be argued my greatest accomplishment was serving as an environmental conscience for the agency. I was the annoying pebble in their shoe reminding them why their management was resulting in a deteriorating planet.

Natural resource managers often insisted their employees pursue and develop proposals that were good for both wildlife and extraction interests. The catchphrase was "win-win situations." It became so prevalent in natural resource agencies that it was a commonly utilized question in job interviews. *What win-win situations have you achieved?*

Win-win situations were hidden in pots of gold being guarded by leprechauns. Numerous fairy tales were created by managers and their staff to rationalize environmental decisions being a win-win. It was the natural resources equivalent of alternative facts. When it comes to economic growth and environmental conservation, there is no such thing as a win-win situation. One always comes at the cost of the other.

The win-win situations that managers gloated over were shell games where the planet always lost. Almost without fail, a single species was spared from being destroyed in a small area in exchange for an area elsewhere being annihilated on a much grander scale.

In the last 30 years, working groups have become an integral part of how the bureaucracy addresses larger environmental or conservation issues. *Consultation, coordination, collaboration,* and *cooperation* are the operative buzzwords. These groups often take on a life of their own. A number of people's careers have been defined by a single working group. Examples are greater sage grouse working groups, prairie dog working groups, and invasive species working groups. At a planetary level, there is the Intergovernmental Panel on Climate Change. While there are sincere people in all these groups, there is neither the agency nor the political will to address significant issues in a meaningful way that might avert the environmental crisis for which they were formed.

With few exceptions, their only value is to document in exquisite detail various trends that portend the collapse of whatever species or issue the working group is attempting to address. We're clever enough to identify the crisis but not nearly smart enough to extricate ourselves from our self-induced predicament.

Over time, these groups independently become experts in the same activity—beating a dead horse. It's often possible to skip an entire decade, walk into one of these meetings, and think you never left. They will be discussing exactly the same issues. The only difference is there will be different people in the room, and whatever issue they were formulated to resolve is now much worse.

Prior to the Agricultural Revolution, at the end of the day, success was being alive and going to bed with a full stomach. It was usually achievable. There were no gods to help them recover from illness, win the lottery, kick the game-winning field goal, or push the wind in a direction where a lion couldn't detect their presence. There was socialism, no justice system, no political system, a different concept of time, no economy, no money, no capitalism, no corporations, no patriarchy, animist religions, and our ancestors were almost certainly happier and more content than we are today.

When humans made the Faustian bargain with wheat and agriculture, success was no longer achievable. Early farmers regretted not plowing, irrigating, and weeding more. They felt guilty about not protecting their crops from wildlife better. The harvest was never as successful as it could have been. Eventually, remorse developed over not having enough land to hire serfs and slaves and becoming ever more powerful. For the last 10,000 years, the power, affluence, and riches people achieve have never been quite abundant enough to feel as successful as they had hoped. Even the most powerful and wealthy are doomed to live with anxiety and anguish, forever chasing after greater pleasures.

Happiness is most tied to our expectations. When things improve, expectations balloon, and consequently even dramatic improvements in objective conditions can leave us dissatisfied. When things deteriorate, expectations shrink, and consequently even a severe illness might leave you pretty much as happy as you were before.

As we aspire to more success, we consume more natural resources. We are consequently wreaking havoc on our fellow plants and animals and on the surrounding ecosystems, seeking little more than our own comfort and amusement yet never finding satisfaction.

Humanity is accelerating to its inevitable endpoint with no ability to pump the brakes, because there are none. Just holding steady costs more than humanity can afford. No one can even slow the growth. Economic growth at the cost of the planet's many life-forms will continue until the planet forecloses on humanity.

CHAPTER 7

THE PSYCHOLOGY OF EXTINCTION

Humanity employs two behavioral mechanisms that blind us to our status on the planet: cross our fingers and hope nothing bad happens, and ignore the obvious.

HOPE

Hope is the elephant in the room. Tempering environmental bad news with hopeful qualifiers has reached an art form. There is no shortage of troublesome environmental news, so writers are well-practiced. Authors strive to have their labors read but innately realize humankind can only handle limited amounts of reality: Environmental problems are dire, to be sure, but with a few changes in lifestyle, humanity can turn the tide, ensuring our own survival while at the same time securing the health of the environment. Happy endings make for marvelous reading.

The reader's needs are consummated by writers with caveats and qualifiers, the most commonly used being *if*. *If* we address climate change by a certain year. *If* Brazil stops deforestation of the Amazon. *If* we do all these things and more, the planet will remain habitable. Other than a few tiny tweaks, we'll be able to continue our lives unaffected. "There is still time!" is another favorite of those writers believing humans will shapeshift into a different creature.

Hope is derived from our most optimistic expectations. Who hasn't hoped to win the lottery? Meet their perfect mate to spend the rest of their life in blissful harmony? Retire rich by the time they are 50? Be a professional athlete, actor, or personality? Have a good day at work? Land a dream job? We have new hopes every day, and our existing ones morph on a daily, even hourly, basis.

Actual circumstances often intercede to modify our hopes and occasionally to crush them. Most people can't recall the number of times in their own lives they have abandoned hope. The most common reaction to broaching the topic of humanity's limited future is "How can you go on living without hope?" These same people overlook that nearly everyone has a long list of experiences where the reality of a situation forced them to give up hope. Rarely does someone stop living or discontinue the pursuit of happiness because of dashed hopes.

When I was a child, my parents made an elaborate charade out of Santa Claus. My sisters and I were chagrined because we had an oil stove rather than a fireplace chimney. How would Santa Claus get into our home to deliver presents? Our parents assured us that Santa was just as adept at negotiating the windows as he was crawling down chimneys. Santa Claus brought my sisters and me a lot of happiness and suspense early in our lives. Somewhere between the ages of eight and 10, the reality of the situation became apparent, and we ultimately gave up hope in the existence of Santa Claus.

Almost all people hope their marriage will last a lifetime. However, approximately 50 percent of all marriages end in divorce. Either one or both partners decide their marriage was hopeless and they would be better off alone or with someone else.

I've had good friends, family, and pets that were injured, lost, or gravely ill. At some point, the reality of the situation required my giving up hope they would survive. Similarly, everyone will become active participants when hope loses its battle with their own mortality.

The disconnect with losing hope for humanity's continued existence is one of scale. It's a case of reality overload with a surcharge. No one gives it a second thought when they don't win the lottery. However, the thought of humanity's finite continued existence is too much to contemplate. Unlike finding another person to marry or purchasing another lottery ticket, extinction is final. As a result, denial is the mechanism most utilized to ignore the distasteful outcome. A dollop of honey disguised as hope is necessary for those capable of dabbling in such an unpleasant topic.

The imminent peril to humanity can only be sugarcoated by spinning a mountain of information suggesting otherwise into sunshine, lollipops, and rainbows. There are no scenarios where humans magically change into a different organism. This book explains the impossibility of changing our current trajectory and preventing the extinction of most complex organisms on the planet, including ourselves. Humanity has already traveled most of the way toward the end terminal and is currently preparing for departure.

The only variable that can be tinkered with is the speed at which we arrive at our destination. There is a very small percentage of the population who either consciously or subconsciously know humanity has a limited future. As the environment continues to deteriorate and the reality of our situation becomes increasingly obvious, the numbers of those recognizing how powerless humanity is to reverse course will continue to grow.

These hyperaware people who recognize our trajectory don't necessarily have an environmental background but come from all walks of life and are of all ages. Because of their practiced ability to compartmentalize, those working in environmental fields have an even greater capacity to ignore humanity's firmly entrenched path than the general population.

DENIAL

Denial is one of the most powerful tools in the human psychological repertoire. Faced with a painful fact, we have an uncanny ability to reject the reality of that fact. The more invested humans become in a myth, the less open they are to alternative ideas that might conflict with that belief. Many people would rather die than confront a gut-wrenching truth, especially if it conflicts with long-held beliefs.

We might compare our behaviors to how ancient Greeks reacted to Cassandra, a Greek mythological prophet. She was condemned by the gods to utter prophecies that were true but that no one believed. When the Greeks appeared to quit their assault of Troy and left a wooden horse outside the city gates, Cassandra warned the celebrating Trojans that there were Greek warriors hidden inside. The Trojans ignored Cassandra and brought the horse into the city gates with tragic results. Any situation can contain truths that aren't readily apparent to most people but are nonetheless visible. Cassandra's experiences also embody the frustration we feel when no one else can see what seems obvious.

The information provided in this book may be most beneficial for all the Cassandras in the world who are insightful, curious, open-minded, and courageous. It is intended to help people who already partially or wholly comprehend our predicament more fully understand why we can't and won't reverse course. Understanding our destiny can make happiness more difficult to achieve but doesn't preclude it.

Concerns for the environment have routinely been dismissed. The economy, foreign policy, substance abuse, discrimination, and a host of other issues are viewed as paramount. As a result of these priorities, politicians can't afford to focus on the environment for fear of being voted out of office. The economy has always been more important than the environment.

Economic growth is attained by increasing the extraction of ecological wealth and despoiling the planet. Most people everywhere want undisturbed ecosystems, clean water and air, and abundant wildlife. These same people also want economic growth. Given the choice, economic growth is always the higher priority.

There are always environmental costs to economic growth. For centuries, those bills have been shuffled to the bottom of the mail stack by humanity, forever unopened. The debts have, thus far, been paid by thousands of life-forms that have quietly winked out of existence without whining or complaining. Also on our growing tab are decimated forests, grasslands, and marshes, and despoiled oceans. The planet is quickly running out of capital to pay the mounting bills. Bankruptcy is imminent.

The inability for me to see for most of my life what now seems obvious is another example of how hope can blind even the most well-trained. Saving ourselves and the biodiversity of the planet is beyond the capability of our evolutionarily programmed behavior.

The more onerous the outcomes, the more extraordinarily difficult they are to accept. Most people expect their lives to turn out better than they wind up being. Almost everyone underestimates their chances of losing their job or being diagnosed with cancer, envisions themselves achieving more than their peers, and overestimates their likely life span, sometimes by quite a lot.

Optimism bias is the belief the future will be much better than the past and present. It's present in every age group, race, region, and religion. Despite the continual barrage of news about rising sea levels, increasing temperatures, pollution, and species extinctions, private optimism in our personal future is incredibly resilient.

Overly positive assumptions can lead to disastrous miscalculations: make us less likely to see a doctor regarding a nagging, chronic ache or pain; less likely to save money for a personal financial crisis; or in this case, more likely to believe everything will turn out fine in spite of a host of calamitous environmental issues.

Optimism may be hardwired by evolution into the human brain. It keeps our minds at ease, lowers stress, and improves physical health. Optimism left me thinking others would eventually see how important preserving planet biodiversity and protecting ecosystems was for our own and the planet's future. Five decades of interacting with people, wildlife, and nature leads to an easy conclusion: a heightened awareness will never happen.

IGNORING

Pliny the Elder, a Roman author and naturalist, may have started the myth of ostriches burying their heads in the ground when he wrote that the birds "imagine when they have thrust their head and neck into a bush, that the whole of their body is concealed." Even though ostriches don't bury their head in the sand, the analogy is apt and oft used to describe people who ignore the truth and refuse to accept the state of the world as it actually exists.

The cognitive limits of our brain simply won't allow us to perceive and know everything. To prevent our senses from being overwhelmed, we filter what we hear, see, smell, and feel. This editing process becomes critical to how we perceive the world.

The information we allow past our filter is mostly what makes us feel good about ourselves. Whatever might make our delicate egos uneasy and conflict with our most vital beliefs is edited out. This seemingly harmless behavior deprives us the ability to understand and perceive the world and our habitat.

Nearly everyone wants their life to matter or have some deeper meaning. There is no shortage of self-help books, inspirational speakers, holistic coaches and healers, and preachers directing our energies so we can happily aspire to greatness. We desperately want to believe our lives are more than the sum of our seconds, days, and years on Earth and a temporary vessel holding molecules that will be recycled over and over again for millions of years. This desperation is the fuel that runs the engines of religion and all manner of activism.

Time perception is a fundamental element of human awareness. Our consciousness, our ability to perceive the world around us, and ultimately, our sense of self are shaped by our perception of time in a loop connecting memories of the past, present sensations, and expectations about the future.

Historians, archaeologists, and paleontologists help us look at time immemorial. However intensely we look, seeing is not an option. Seeing reduces our lives to a meaningless speck in the universe. If complex life on Earth could be encapsulated by a 24-hour day, someone who lives 90 years would be represented by 1/100 of a second. The entire history of the United States is 4/100 of a second. Humans have destroyed most of the ancient forests and grasslands on the planet in 3/100 of a second.

Christianity has been in existence for about 40/100 of a second. The Agricultural Revolution began 1.7 seconds ago. The sixth extinction on the planet began in earnest about eight seconds ago. Humanoids have been present for 13 minutes. Saber-toothed tigers were top-tier predators for an hour and 15 minutes. The dinosaurs managed to clock in a full eight-hour shift. The period of time between when early hominids branched off from the great apes to today's modern humans wouldn't make it through the dinosaur's coffee break.

To feel like our lives are meaningful, we focus on the present to the exclusion of the past and future. No one wants to admit the people in their lives who died 20 years ago are forgotten by all but a few. People who died 40 years ago are remembered by even fewer. People who died 60 years ago are rarely remembered except in the most abstract of ways. The vast majority of those who died over 80 years ago are forgotten. A few people might know they existed, but they quickly lose their importance. Those who are now dead and forgotten almost certainly had the same aspirations for a meaningful life as we do today.

By only acknowledging our lifetime, concerns about the effects of our actions on the planet are easier to ignore. The irony is while what happened 40 years ago or what will happen 40 years from now aren't meaningful to most, it doesn't prevent us from sincerely believing that our significance on the planet will somehow transcend those who came before us. We stubbornly ignore time outside of our immediate experience, unwilling and unable to see.

Most people aspire to preserve an image of themselves as stable, accomplished, and virtuous. These beliefs are a vital and central part of not only our self-image, but how we are perceived in the eyes of our friends, family, and colleagues. Anything or anyone who dares question how we perceive ourselves produces suffering that feels just as dangerous and unpleasant as illness or thirst. A challenge to our big ideas makes us feel as if we are in imminent peril. We ignore evidence that proves we are wrong or reinterpret evidence to support our beliefs to minimize or alleviate the pain. In essence, our egos are doing the same thing as the Putin and Kim Jong-un totalitarian regimes: controlling information.

To support our self-image, we conveniently overlook or ignore threatening or incompatible ideas and evidence and even rescript our own history. To ensure we don't destroy our sense of who we are in the world, we reinterpret events to fit our expectations.

Most people are guilty of confirmation bias, the tendency to look for proof of our own assumptions. We only believe what we want to believe. We only know what we want to know. Many people who don't want to believe something won't believe it until we see it and feel its impact on ourselves. People readily build houses in known flood plains or in forested areas, despite the risks of floods and fires. People who weren't personally affected by the coronavirus often felt it was abstract and not relevant to them, until they were directly affected. Humanity's cognitive flaw of confirmation bias makes dealing with climate change and the loss of biodiversity impossible. There isn't a large enough percentage of the global population willing to challenge our own biases to make anything but superficial efforts to address catastrophic environmental issues.

We are the only known self-gaslighting species. When we get attached to a theory, we embrace it like a toddler with a teddy bear. We screen out unhelpful facts, invent favorable ones, and ignore contradictions in our own claims. When people become motivated enough to believe something, they are going to believe it no matter what. There is no such thing as a bridge too far.

The entire framework of our existence is the result of myths created by humans since the beginning of the Agricultural Revolution. Myths are reinforced by gathering around us the people that reinforce our beliefs and ideas. Without the support of friends, family, and colleagues, it is more difficult to do things and easier for questions concerning our ideas and beliefs to creep through our carefully constructed filters. The social support allows us to act on the big ideas that bring like-minded people together.

If you are in a position of tremendous institutional or political power, then not only are you hugely confirmed by the followers who share your beliefs but those who dare question them would threaten everything: identity, social status, reputation, and future career. We're all operating in a dense fog of mutual reinforcement. A deep flaw of humanity is mistaking agreement for truth.

The more insecure a leader or manager, the less their tolerance for employees who rock the boat and the more they surround themselves with sycophants. This is especially true in large bureaucracies and corporations. I was well-known for not being a team player in the bureaucracy, twice being sent to what was affectionally referred to by colleagues as "charm school."

The good team player in the wheels of environmental-agency bureaucracy is implicitly viewed as the person who goes along with the team, not the one who asks hard questions. When I would relate my experiences to the instructors at my punishment trainings, they would shake their heads in disbelief.

What they actually taught was teams are established to solve problems, not unconditionally support a preconceived outcome by superiors. Being a truly good team player involves having the confidence to dissent. Paradoxically, the trainings I was sent to as a punishment were the only times when my actions were considered completely reasonable. Only groupthink was condoned when I returned to the office.

Government agencies and large businesses provide the perfect breeding ground for the creation of chronic fawners, toadies, and flatterers reinforcing whatever worldview their leaders espouse. An instinctive bias in favor of one's "in group" and its worldview is deeply ingrained in human psychology. Accepting a group's ideological belief system, whether grounded in science or illusion, is necessary to truly belong. This is true in work, family, and most aspects of life.

When ideas are widely held, they don't stand out as much; they can even become the norm. As long as everyone believes climate change, deforestation, desertification, and the loss of biodiversity won't affect our lifestyle, we can continue on with our lives unaffected. A widespread faith in this fiction feeds into our predisposition for mistaking agreement for truth.

Humans often prefer ignorance to knowledge. One of our primary tools for dealing with conflict and change is to imagine it out of existence. A preference for the status quo, combined with an aversion to disagreement, compels us to turn a blind eye to problems and conflicts. For example: avoiding having a serious discussion with a coworker or significant other before a conflict spirals out of control; procrastinating starting a term paper; knowing our food choices or drinking habits are unhealthy. We often think and hope that if we ignore these uncomfortable situations, they will go away. We cling bravely to the commonplace.

People who don't believe in evolution walk into doctors' offices every day and readily accept medical treatments whose framework is based on the science of evolution. Sometimes, the doctors providing the treatments are ensnared in the same cognitive dissonance as their patients.

Avoidance is the number one factor that keeps people stuck in trauma response and PTSD. Traumas and trauma reminders trigger our autonomic defenses: attach, submit, freeze, fight, and flight. Too much "autonomic noise" often results in people disregarding worries and traumas.

Our imminent extinction may be creating an anticipatory traumatic experience for some. Most people have a basic understanding of the fight, flight, freeze, and submit autonomic defenses. However, attach may be the least understood, most well developed, and universally utilized by humans. It is our first line of protection as infants. When a baby perceives they aren't getting proper care from a primary caregiver because of lack of attentiveness or availability, they cry and try to attach to them so their needs are properly met, increasing their chance of survival.

As people age, the attach defense is utilized in more sophisticated ways. If, for example, a significant other doesn't seem as available or seems to be pulling away, instead of crying out like a baby, an individual might obsessively long for them, check their phone every five minutes to see if they've called, or send them a flurry of messages every hour. When dealing with our deteriorating planet, this defense mechanism can manifest itself by accepting whatever the media says, consumerism, or blind political activism.

Window of tolerance is a phrase used to describe when people are typically able to readily receive, process, and integrate information and respond to the demands of everyday life. During times of extreme stress, people often experience periods of either hyper- or hypo-arousal.

Hyperarousal triggers a fight-flight response and is often characterized by hypervigilance, feelings of anxiety and/or panic, and racing thoughts. Hypo-arousal, or a freeze response, may cause feelings of emotional numbness, emptiness, or paralysis. In either of these states, chaotic or overly rigid responses become the norm and the person is said to be outside the window of tolerance. Our vexing situation may be activating many people's autonomic defenses, making *Homo sapiens* incapable of processing our dire circumstances, however obvious they may be.

Too much bad news has people either looking for myths to rationalize it in a positive way or pretending it doesn't exist. Just in the last 30 years, people no longer die—they pass. Slaughterhouses no longer exist. Instead, they have been magically transformed into beef and pork production facilities. Creative linguistics has now made death less final and more palatable. Similarly, people don't want to hear about an imminent catastrophic collapse of the planet or civilization. Accepting that reality would conflict with the carefully constructed story about both themselves and their world.

Research has shown that even a small amount of negative brain activity can lead to a weakened immune system, making a person more prone to illness, and even lead to a heart attack or a stroke. Negative attitudes can also affect the ability to think clearly, compromising the effectiveness of neurons in the hippocampus. This area of the brain is responsible for reasoning and memory.

Oprah Winfrey and many other motivational speakers in the world advocate surrounding yourself only with positive people who are going to inspire you and lift you higher. Not only are we physically hardwired to avoid negative thoughts and situations for our health, society helps us do the same.

To make ourselves happy and feel good, most people have lost the ability to discern between those who have very real bad news and those who are curmudgeons. They are all captured by the same internal filter protecting our sense of self. Even more dangerous are the charlatans described in the next chapter espousing good news that pour through our filters unimpeded.

It is now possible to see warnings of our imminent demise in the news every day. However, these warnings are usually tempered by hope. Scientists don't want to face the reality any more than other people. They are just as human as everyone else and have the same behavioral mechanisms for avoiding uncomfortable information as the public at large. They also rely on funding for their research. What institution will fund a scientist who proclaims humanity is beyond hope? What advertiser will promote a product on a news show that reveals to their clientele that *Homo sapiens* has a very limited future?

The effort to sugarcoat warnings is often done with the best of intentions. Unfortunately, trying to reach what is a largely willfully blind public is ineffective for no other reason than it is impossible. As I did with endangered species conservation my entire career, most do so hoping the public will take notice and make taking care of the planet a bigger issue than the economy, immigration, discrimination, or substance abuse. All of these are real issues today, but when there are no humans, they cease to exist. Focusing on them to the exclusion of paying attention to the environment is like a homeowner worrying only about a leaky faucet when their roof is about to collapse.

So, how can we largely ignore the overwhelming number of environmental issues? By not acknowledging the problems, most people hope and believe they will go away. It's a uniquely human attribute. The gravitational pull of the status quo is incredibly strong. It feels easier and less risky, and it requires less mental and emotional energy to go with the flow. Familiar habits and beliefs feel safer because we're used to it. It's so much easier to imagine that what we don't know won't hurt us.

A decade ago, I was visiting a friend in California. It was the third year in a row of drought in the state and the first in California's history for which a statewide proclamation of emergency was issued. My friend and I were discussing climate change and other environmental concerns when he asked me how long before there would be a crisis. As it turns out, I may have learned more from that single question than any other in my life.

My reply to his question was five years. As it turns out, it was both the right and the wrong answer. The first thing I learned is a precise scenario for our future demise is impossible to accurately predict because chaos and random events are such a big part of the equation. So many forces are at work, and their interactions are so complex that extremely small variations in the strength of the forces and the way they interact produce huge differences in the way things will play out. Despite the unpredictability of potential scenarios, because there are so many negative forces in play, the outcome is easy to predict even though its exact time frame is not.

More importantly I learned for most people, no matter what happens, there will never be a crisis. Crisis is defined as a time of intense difficulty, trouble, or danger. Rarely does someone recognize a crisis until they are already in the middle of it. It is likely few people directly affected or killed in the wildfires in Spain and Portugal in 2017, California in 2018 and 2020, Australia in 2019 and 2020, or Hawaii and Canada in 2023 thought global warming was a crisis. Most of the people who narrowly escaped with their lives or lost their homes probably still don't think global warming is a crisis. The crisis was the wildfire. Because global warming always exists and is not present at any one place and time, it becomes less concrete in people's minds and less capable of being perceived a crisis.

Without a doubt, the wildfires weren't a crisis for the people living in the surrounding communities. The smoke was annoying but could hardly be termed a crisis. It certainly wasn't a crisis for the rest of the people in the country or on the planet. In essence, it wasn't a crisis for anyone but the people who were killed, and it was only a crisis for them moments before they died. They probably didn't even have time to recognize it as a crisis.

Crisis only becomes a crisis when it directly affects people themselves, their close friends, or their family. After any major failure or calamity, voices always emerge saying they'd seen the danger and were warned about the risk but the warning had gone unheeded. There were a number of people warning the 2008 financial crisis was imminent prior to it happening. I sold my home in 2007 and luckily chose to rent an apartment rather than buy another home. As a result, the financial crisis of 2008 wasn't a crisis for me. It was only a crisis for those people who did purchase homes or had invested heavily in the stock market.

Climate change was a crisis for those people living on islands who have been forced from their homes by rising sea levels. Climate change will be a crisis for all those people living in coastal communities when they will likewise be forced from their homes. However, climate change for most of the population on Earth will never be perceived a crisis in spite of nearly everyone being negatively affected in a myriad of ways.

People will increasingly perish as a result of our incursions on the planet, but it's highly unlikely that most will believe it is a crisis until too late, if ever. Even fewer will make the link between their crisis and humanity's continued assault on the planet's natural resources.

Because of all the layers of insulation we've created between ourselves and the planet, few understand that environmental issues are often the underlying cause of almost all modern crises. Most people can't and won't make the connection that each event was made much worse by an environmental problem. Ignoring the complexity simplifies their lives and allows them to continue on a path pursuing the status quo for as long as possible.

ROMANTICIZING

Humans reconstruct the past by picking up little pieces and putting them back together in a pleasing way prone to various biases. We carry our past with us and are subject to retrieval-induced forgetting. The aspects of our ancestors we enjoy thinking about stick with us over time, creating a nostalgic haze. We start out with the hypothesis that things were better in the past and then utilize confirmation bias to affirm that.

People tell themselves that surely there was a time when humans weren't the destructive denizens of the environment we are today. We want to create a coherent identity and favorable sense of our species over time. We often imagine ways in which the past could have happened and then use our imaginations to further modify the content. It's an adaptive way to regulate emotion in the present and enhance optimism about the future. Once an evolutionary period in our history is over, it is easy to blur out any imperfections and bathe it in a hazy, soft-focus glow.

A romanticized version of the past is fun. We have a long history of glossing over the past, choosing to overlook the darker, grittier, harder-to-address parts of it. Only by viewing history wholly and objectively can we understand how humanity arrived at our current status on the planet.

Romanticizing our ancestors has created a long-running tenet that humans evolved as hunters. Being a hunter-gatherer is so much more glamorous than being a scavenger-gatherer or thief-gatherer. Humanity loves to equate themselves with eagles, not buzzards. Leopards, not jackals. The mostly fictional sentiment of humans as hunters was also fueled by the Judeo-Christian ideology of humans being inherently evil and dangerous. The reality is humans evolved as a prey species. Human cooperation may have partially arisen because of group efforts to outwit predators.

Many ancient cultures and peoples are perceived or portrayed as living in harmony with nature and each other. Seeing their knowledge as a fixed and stable system is an illusion. People were just people and were not all the same. They changed and adapted over time, and their knowledge evolved based on local conditions. Romanticizing people leads many away from the reality and truth of the world, creating fairy-tale stereotypes of our ancestors, both recent and ancient. It is usually the result of looking at a specific group of people and freezing them in a place and time. The same can be done with individuals in modern society. These are always exceptions, but evolutionary processes occur most often when certain traits or behaviors provide a short-term advantage to individuals within a population, increasing their chances of survival and reproduction. Those individuals and groups of people throughout history without a narcissistic short-term self-interest are so few that they have no significance in evolutionary time. Believing exceptions are the rule is one more way of tricking ourselves into believing humanity is capable of something "more."

Many believe *Homo sapiens* is unique and remarkable because of our capacity to experiment with alternative forms of social organization, our capacity for self-creation, and to collectively contribute to decisions about how to live together. For most people, rationalizing our "specialness," whether currently or historically, comes as naturally as breathing.

To romanticize something means to depict it in an idealized or unrealistic manner, with the intention of making it more attractive than it is. This tendency often leads to continual disappointment in modern humans, as they fail to live up to the idealized image of our past selves. Moreover, it fosters an unrealistic hope that humanity can somehow transform into something greater. The unknown scares us. We feel anxious and uneasy when we can't figure out how things will be like in the future for us. Romanticizing our past provides an escape ramp from the path to a fearful unknown.

ACCEPTANCE

Recognizing and accepting our situation is a double-edged sword. I often envy people who are naive to humanity's future. It would certainly be nice if the most important issues were whether we can meet a critical deadline at work, whose sports team will fare better next year, what movie is playing at the theater, or where to vacation.

Acceptance of our imminent extinction is incredibly difficult. Regardless, the physical and mental tools necessary to attain happiness aren't any different whether a person is completely naive or not. There are many people with no understanding of the ramifications of our planetary crisis that still suffer from depression and other behavioral maladies.

People live in a dual reality. There is the objective reality of wildlife, trees, rivers, and oceans. There is also the human-designed reality of gods, nations, and corporations. In today's modern world, more and more people are completely divorced from the objective reality. They might know it is out there somewhere in a vague way, but it is not part of their personal experience.

No one wants to acknowledge that our lifestyles are killing us. The dissonance produced by reading about our environmental impact on the one hand and continuing to live as we do on the other hand is resolved by minor alterations in what we buy or eat, but in very few significant social shifts. Even people who believe there are environmental problems rationalize away their own contribution to destroying the planet by buying a hybrid vehicle, voting Democrat, becoming a vegetarian, or donating money to an environmental group. The gravitational pull of the status quo exerts its overwhelming influence.

In the meantime, global conferences end when no one has the stomach for the levels of disagreement they cause. No politician shows the nerve for the political battles real change would require because they know it is a battle they can't win. Their constituents would vote them out of office.

We conform to the consumption patterns we see around us as we all become bystanders, hoping someone else somewhere will intervene. Our governments and corporations have too many different and connected parts to communicate or to change. Our only consolation is the money we chase with reckless abandon.

During my career, I was often left trying to answer the question: Why doesn't humanity stop what they are doing? Regularly witnessing individuals destroying the environment, even on the public's land, for personal gain was sad and frustrating. Society ultimately paid and is continuing to pay the price, occasionally willingly, but usually unknowingly. Ultimately, the price humanity pays will be exorbitant.

Most people are too far removed from the planet that supports us or clinging desperately to the status quo to acknowledge our role in destroying it. Humanity will disappear with only a relatively few people acknowledging or even aware of our starring role in the tale of horror. As life on the planet increasingly unravels, finger-pointing, recriminations, selective amnesia, and violence will become the norm. It's who we are and what we do best. It is already happening.

CHAPTER 8

DECEPTION

People who are unfamiliar with the natural world and whose only exposure is what they read or see on television become susceptible to con artists disguised as environmentalists, scientists, or self-proclaimed experts. These hucksters often function as puppets of industry and extraction interests. Even knowledgeable people can, at least briefly, be hoodwinked by these clever wolves in sheep's clothing.

Hiding behind titles, awards, and degrees, their capability for facilitating extraordinary environmental destruction through a campaign of misinformation outstrips overt anti-environmentalists by a wide margin. These charismatic fraudsters are jumping on the accelerator of the bus speeding humanity over the cliff. Their expertise lies in taking a few truths, ignoring others, and spinning fictional narratives that people want to believe. In their efforts to win elections, many politicians effectively spoon-feed people a tale they want to hear and trust. However, politicians could take lessons from these masters of smoke and mirrors.

Stephen Mealey began his career with the Forest Service in 1977 as a wildlife biologist in Cody, Wyoming. As is often the case with quick promotions in natural resource agencies, Mealey became a planning specialist elsewhere in 1980. Planners can easily avoid agency strife because they spend much of their time in an office writing plans or going to meetings. Planning can be a fast track to promotion for natural resource agency employees who prioritize career over natural resources. It was no different for Mealey, who returned to Cody as the forest supervisor in 1983.

Under the pro-logging Reagan administration, when Forest Service supervisors were judged by how much timber their forest cut and sold, he quickly became a rising star. After a stint in Washington, DC, in late 1991, he became the forest supervisor of the Boise National Forest.

Mealey was the first in the agency to effectively weaponize the term *forest health* as a means to increase timber harvest and log old-growth forests. Much of the forest was overstocked with small trees as a result of historic timber harvests, overgrazing, and fire suppression. Reducing these thick stands was a legitimate need to protect the old-growth ponderosa pine trees interspersed across the landscape. Mealey used the need to remove the small trees as rationale to obscure his primary goal of logging most of the forest's remaining old-growth ponderosa pine. In so doing, he provided the agency with a blueprint for cutting ancient trees under the guise of forest health that is still utilized today.

During the three-year period Mealey was the Boise Forest Supervisor, he increased the forest's timber harvest by 28 percent. To put that in perspective, during the same period, the entire agency ratcheted back timber harvesting 38 percent. His ability to increase timber harvest at a time the Forest Service was attempting to take their mandate to conserve ecosystems more seriously ingratiated him with the governor and timber lobby in Idaho. There was nothing that played better with an unknowing public than a former wildlife biologist who could increase timber harvest while espousing forest health.

To further take advantage of Mealey's lack of a resource ethic and maximize future timber harvesting, the Idaho governor and his commission selected him as their director of the Department of Fish and Game. This was despite learning he had been convicted of poaching a deer when he was younger. Mealey may have the distinction of being the only director of a state fish and game department in American history to have been convicted of poaching a big-game animal. While Mealey was certainly not alone in using his college degree and title as props to undermine the planet, he was one of the most effective, with a legacy that has lasted long after his retirement.

One morning while browsing the internet, I happened on an article titled "Why Apocalyptic Claims About Climate Change Are Wrong" from *Forbes* magazine. I thought, *Great! Finally! Someone who can prove me wrong.* The author was a man named Michael Shellenberger. He had all the awards and qualifications to make anyone immediately believe he would make a particularly legitimate and compelling case for his assertion. He was nationally recognized as *Time* magazine's "Hero of the Environment" and a Green Book Award winner and founder and president of Environmental Progress.

He started out by claiming that over the last four years, his organization, Environmental Progress, had worked with some of the world's leading climate scientists to prevent carbon emissions from rising. Ironically, on the same day his article was posted on the internet, the latest State of the Climate in 2018 report revealed atmospheric carbon dioxide at 407.4 parts per million (ppm), a new record high and 10.2 ppm higher than in 2014.

Laughably, in the next sentence, after his immediately debunked boast, he insisted on getting the facts and science right by correcting inaccurate and apocalyptic news media coverage and criticizing those who didn't share his distorted worldview. Shellenberger then asserted that no credible scientific body has ever said climate change threatens the collapse of civilization, much less the extinction of the human species.

While technically right, he seemed to believe climate change will work in a vacuum, completely independent of every other insult humanity is inflicting on the planet. Although a self-proclaimed expert, he repeatedly examined climate change through the wrong end of the telescope. According to Shellenberger, the only way people might die from climate change is from climate-induced disasters. Species extinctions and catastrophic disasters have a consistent theme. They are rarely, if ever, attributable to a single cause. Dealing with either requires considering and dealing with a multitude of complex variables and contributing factors.

The numerous insults on the planet that humanity has created over the last 100 thousand years have achieved the capacity to greatly exacerbate any natural or human-caused event. We're seeing this around the planet with more and more frequency. Global warming didn't start the wildfires in Spain, Portugal, the United States, Canada, and Australia, but thanks to climate change, instead of inconsequential wildfires, they became roaring infernos. Hurricanes have always happened, but climate change is making their effect much more significant. Most people don't understand that human population density, climate change, and the loss of biodiversity were the likely drivers of coronavirus becoming a pandemic. A loss in biodiversity usually results in a few species replacing many. These few species tend to be the ones hosting pathogens that can spread to humans and remaining animals. Rather than a loss of biodiversity, the problem is a highly contagious virus and a lack of commitment to wearing masks and social distancing.

For most people, including Shellenberger, those were wildfire, hurricane, and virus problems. When peering through the small keyhole of their existence, climate change, biodiversity loss, or population density is nowhere in sight.

The medical field provides the most apt analogies as to how people consistently view environmental crisis. A person dies of heart disease, not of a lack of exercise and eating at fast food restaurants for a decade. People die from cancer, not increased toxins in our environment. Symptoms, rather than causes, are always easier to deal with psychologically. Causes are usually big, chaotic, and messy. They complicate our worldview.

Few species can withstand a series of concentrated simultaneous stresses on a variety of fronts over a number of years. When all the natural resources supporting an organism are stretched to their limits, it takes only a little nudge in any number of ways to precipitate catastrophic scenarios that result in a species' extinction.

Few people understand how natural resources and ecosystems support our lifestyles. This results in the inability to recognize the deterioration of our planet as either the root cause or, more commonly, a major contributor to calamities occurring with increasing frequency.

In his article, Shellenberger made the same tired argument that Mealey and the timber industry have made for the last four decades. Timber industry hacks seek to improve perceived forest health problems created by 150 years of logging with more logging, and Shellenberger's answer parroted the same argument. Improving the condition of our planet reeling from the effects of economic development in the Industrial Age is with more economic development. He's every corporate giant's dream mouthpiece come true.

Shellenberger claimed to be concerned about the impact climate change will have on endangered species. Whether profoundly ignorant or monstrously disingenuous, he ignored the uncomfortable fact the planet is currently experiencing an extinction crisis mostly stemming from exploitation of natural resources resulting from humanity worshipping at the pulpit of consumerism.

Three hundred thirty-eight land vertebrates have been documented going extinct since the year 1500. An additional 279 species have gone extinct in the wild or are possibly extinct. Of those 617 species extinctions, 77 percent have occurred since 1900. Fish, invertebrates, and plants would throw the number of extinctions during this period into the tens and hundreds of thousands.

Climate change had almost no effect on the long list of extinctions that have occurred up until this point. Only in the last few years has global warming thrown its hat in the ring in becoming one more significant threat to the existence of native organisms on the planet. Until recently, habitat loss driven by economic development, not climate change, is the most significant factor resulting in the current deteriorating status of native fish, plant, and wildlife populations and the planet.

Climate change isn't a rabbit the magician pulled out of a top hat. Prosperity and economic growth fueled by industrialization and deforestation created climate change. Global warming is now having an ever-increasing multiplier effect on the already significant and numerous threats that native plants and animals are experiencing.

Shellenberger's narrative is similar to the generic argument used by devout proponents of consumerism to paint those concerned about the planet's future as irrational. Here is the fictional, but effective, narrative I've heard espoused repeatedly over the last 40 years by economic-growth ideologues:

In the 1970s, Americans were told they were in a global cooling crisis and if something wasn't done, we'd enter a new ice age. When that didn't happen, a few decades later, we were told that entire nations could be wiped off the face of this Earth by rising sea levels if the global warming trends weren't reversed by the year 2000. Despite the consistent failure of these apocalyptic warnings, that hasn't stopped climate change alarmism. We're now being told we only have 12 years to combat climate change, and the solution is to fundamentally dismantle the system of free enterprise.

This argument is akin to a tyrannosaur telling a velociraptor 50 years after the meteor impact, "All of our tribe has been making apocalyptic warnings about our imminent demise as a result of the meteor. Only a few of us have died. Can you believe the consistent failure and alarmism? We're still here, by golly, 50 years later."

Even though dinosaurs in the immediate vicinity of the asteroid impact and lava flows were probably killed immediately, most probably gradually died out over the following decades, centuries, and millennia.

The perturbations humanity has inflicted on the planet are already killing people in ever-increasing ways. Sometimes deaths can be directly attributed to an event, but usually they occur indirectly by exacerbating weather and fire events or health problems. Catastrophic events will increase in frequency and intensity. Dismantling our system of global economic growth isn't an option. For that reason, there is only one viable, appalling option left on the table: the disappearance of humanity from the planet. There are animals poised to survive our disappearance just like there were those that survived the asteroid and volcanic activity 66 million years ago. In fact, not all dinosaurs died. Crocodiles, alligators, birds, and lizards are among those creatures that survived.

The next part of the "don't worry, be happy" narrative goes like this:

The question is even if we believe climate scientists' alarmist, catastrophic predictions, would their proposals even work? Not according to their own models. Based on those models, even if the US cut its carbon dioxide emissions to zero, it would only avert global warming by a few tenths of a degree Celsius in 80 years. We would see no noticeable difference in the climate, yet it would come at enormous costs to the American people. Climate change is happening, and human activity undoubtedly plays a role, but big-government climate policies are all economic pain, no environmental gain.

This part of their argument is vague and muddled because of not defining what is meant by averting global warming. Most climate scientists would be jumping for joy if we arrested carbon dioxide at current levels and we saw no change in the climate. While current levels are having profound effects on the planet, the alternative of steadily increasing carbon dioxide levels are ominous.

Continuing with their well-polished narrative:

Abundant energy sources such as oil, coal, and natural gas have allowed Americans to affordably drive to their jobs, light and keep their homes, and power their refrigerators, computers, and iPhones. More heavy-handed climate regulations would drive up electricity bills and prices at the pump. Families would be hurt multiple times overpaying not just more for energy but also more for food, clothing, and health care, as energy is critical for every stage of planting, harvesting, manufacturing, and transporting goods to consumers.

This part of the argument makes the compelling case that human extinction is preferable to an economic recession. Their argument then takes a nasty turn from silly to ludicrous:

These rising costs would stifle economic growth, one of the most important factors for maintaining a cleaner environment. As a country's economy grows, the financial ability of its citizens to take care of the environment grows, too. So, creating more economy-killing climate regulations and taxes would not only harm the livelihood of the American people but would also harm our ability to protect our environment. Instead, government should focus on keeping the economy strong by reducing taxes and eliminating regulatory barriers to energy innovation. Once these immediate concerns are addressed, we can then concentrate on the environment.

The global economy and energy innovation has never grown more or faster since the beginning of the Industrial Revolution. The environment has never deteriorated more or faster than during that same period. Consumerism ideologues seek to convince people economic growth is a magical process completely independent of environmental deterioration. The reality is economic growth is the most direct and significant factor influencing environmental decline.

Usually, this consumerist narrative is said confidently and concludes with the statement that "it is just common sense." This ensures those who might disagree appear irrational. It's an effective strategy and has worked to convince most people for decades. Everyone believes they have common sense and those who disagree with them don't.

This argument for consumerism resonates with most people because money is the greatest conqueror in history. It turns individuals in every society into ardent disciples. People who do not believe in the same god or obey the same president or prime minister are more than willing to use the same money.

Consumerism has become its own religion. While religions like Christianity, Buddhism, and Confucianism have ideals nearly impossible to live up to, most people today successfully abide by the consumerist ideal. The new ethic promises paradise on condition the rich remain greedy and spend their time making more money and that the masses give free rein to their cravings and passions—and buy more and more. We know we'll really achieve nirvana in return because it says so on all the television commercials. The media supports consumerism because it supports the media.

Economic growth purports to solve most of our ethical dilemmas: marriage problems, poverty, depression, famine, religious fundamentalism, authoritarianism, and a swiftly deteriorating environment. Modernity is based on the firm belief that economic growth is not only possible but essential.

This dogma can be encapsulated in one simple idea: if you have a problem, you probably need more stuff, and to have more stuff, you must produce more of it. Economic growth has thus become the crucial juncture where almost all modern religions, ideologies, and movements meet. The credo of "more stuff" accordingly urges individuals, firms, and governments to disregard anything that might hamper economic growth, ensuring a complete disregard for other organisms and the planet.

In the world today, international connections and interdependence lessens the chance of any one country singlehandedly starting a war. However, it makes all countries more vulnerable to the environmental degradation inspired by those most married to consumerism. Every country in the world is affected by energy policies of China and the United States. The same holds true for deforestation of ancient forests anywhere in the world where they exist. While the new global empire has helped discourage war, it encourages environmental degradation through economic growth. Economic growth always relies on more growth. A stagnant economy cannot survive. The continuous growth of the modern economy without catastrophic consequences is a fraud on a colossal scale.

Economic growth is the supreme good. Consumerism espouses that justice, freedom, happiness, scientific research, and the environment all depend on economic growth. This belief in perpetual growth flies in the face of almost everything we know about the planet and universe. We apply the concept to natural resources and are completely befuddled when seemingly inexhaustible resources disappear. What happened to the passenger pigeon? Rocky Mountain locust? Old-growth forests? Rather than learning from the past, we forge on with our consumerist credo, believing that clean water and air are also inexhaustible. The craving to increase profits and production blinds people to anything that might stand in the way. In a competitive economy, it is in every individual's and country's interest to exhaust available natural resources in the pursuit of private and national wealth.

Humans have and will continue to cut down forests, drain wetlands, dam rivers, plow grasslands, and build skyscraping metropolises. Habitats are destroyed and species go extinct. Our once green and blue planet is becoming a concrete and plastic shopping center.

Not only do we not have to cut back on logging to conserve forest ecosystems, but we can also increase timber harvest and make them healthy. Not only will climate change not be a life-threatening issue but more economic development will help solve the problem.

Humans are flooded by impossible amounts of ideas, information, promises, and threats. This has resulted in people relinquishing authority to the free market, to crowd wisdom, and to politicians and other hucksters who tell them what they want to hear. The hand of the market is busy digging the grave of *Homo sapiens*. Hospice is a small price to pay for a little more bling.

While Shellenberger's success mirrors that of Stephen Mealey, he operates at a much larger scale. Both have carefully cherry-picked their facts, distorted others, made misleading comments, and told a rapt audience exactly what they hope to hear. Shellenberger's takeaway message: "Happily, there is plenty of middle ground between climate apocalypse and climate denial." Shellenberger and Mealey are two of humanity's bus drivers glee-fully exceeding the speed limit on the way to the precipice of our existence.

There are many Shellenbergers and Mealeys in the world, although these two have proven themselves effective beyond others' wildest dreams. Snake oil salesmen will always be more popular and able to reach a much larger audience with their pleasant deceptions and half-truths than those delivering a troublesome message based on reality.

Their narratives sound perfectly reasonable to someone whose world-view is limited to concrete jungles, offices, and urban areas. Unfortunately, it is no different than encouraging people to reinforce their roofs so Santa Claus and his reindeer have a suitable landing spot next Christmas.

Those and similar themes play much better than reality and always will because people fervently want to believe them. We're a Pollyanna species singing "Kumbaya" around the campfire as the planet disintegrates around us. Like Shellenberger himself, most people live in a myopic bubble com-pletely independent of the planet that supports us. As a result, few will recognize our demise as environmentally driven.

In theory, any organism can take over the Earth just by reproducing. For example, with just one E. coli bacterium, through normal reproduc-tion, there could be enough bacteria to cover Earth with a one-foot layer in just 36 hours. In reality, this doesn't happen because all organisms have specific needs such as nutrients and suitable environments that are limited. A population can only reach a size that matches the availability of resources in its local environment. In nature, populations may grow exponentially for a time, but they will ultimately be limited by resource availability.

As a population increases, the competition for available resources in-tensifies. It is the biological equivalent of the economic theory of supply and demand. Exponential growth may happen for a while, but when the number of individuals gets large enough, resources start to get used up. The accumulation of waste products can also reduce an environment's carrying capacity. When resources are completely depleted or waste products reach exceptional levels, populations die off. A population crash often follows a peak and can lead to eventual extinction when the environment is unable to support the species any longer. The causes of a population crash can be extremely varied but include a scarcity of environmental resources that are necessary for survival, growth, and reproduction and waste accumulation.

Competition for decreasing natural resources by a rapidly growing human population and the buildup of wastes and toxins are now issues playing out in a wide variety of ways and are described in greater detail in chapter nine. Nations compete with other countries for limited resources. Individuals within countries do the same. The accumulation of waste products in the form of industrial and vehicle emissions, toxins, trash, and chemicals in the environment are quickly reducing the quality and quantity of available clean water and compromising the planet's air quality. Competition for resources has already significantly reduced the global biodiversity that has supported human existence on the planet.

Humanity has been ingenious about making more resources available through migration, agriculture, medical advances, and communication. We've also been creative about using a lot of natural resources when they are abundant and using them all when they are scarce. All these actions have pushed back against our natural environmental sideboards, making it possible for us to exceed our natural carrying capacity. The long-term effect of this short-term success will precipitate a population crash, rather than a decline.

This fleeting success has powered the belief humans are immune from natural biological processes. It has now become an unfathomable concept that human beings are subject to the same planetary limitations as all species. Unfortunately, the only real difference between humans and other organisms is the scale of our activities.

Humanity has not handled success well. Our meteoric ascendancy to the top of the food chain and exponential growth may be at least partially responsible for our complete and utter failure in this regard.

Success is usually fostered by a balance of arrogance and humility. Star athletes and topflight executives are examples. They have the self-confidence to impart their skills and knowledge at the appropriate time and place. They also have the humility to recognize without teammates, coaches, family, and employees they wouldn't be successful. They usually do everything possible to make those around them better knowing their long-term success is contingent on the efforts of those that provide support. Short-term success without humility leads to long-term failure.

What humanity is doing would be similar to a star athlete not only not recognizing their teammates' and coaches' roles in helping them but killing them. The organisms on the planet that have helped foster our short-term success have been discarded like so much trash. Simultaneously, we're using Earth as a garbage dump believing no amount of toxins and trash can harm us.

Self-confidence combined with compassion, gratitude, and a grounding of our self-worth in the context of the planet's ecosystem may have allowed humanity to persist for thousands, possibly millions, of years more. Instead, we threw humility out the window, opting instead for ego, self-aggrandizement, accomplishments, power, and appearance. The evolutionary path humanity willingly walks leads nowhere, quite literally.

Our rapid population growth and the proliferation of wastes and toxins make it highly likely the inevitable population crash will be sudden and intense. Hundreds of millions, and potentially billions, of people will die in a relatively short period of time. Storylines depicted in *Mad Max* movies, while creative, don't seem plausible. A few people persevering in a desolate environment for an extended period is more fantasy than reality.

The most probable umbrella scenario is a one- or two-decade time frame from the beginning of the human population crash to extinction, precipitated by the occurrence of multiple large-scale disasters. There are thousands of specific scenarios where different threats interface with each other in a multitude of ways resulting in the extinction of *Homo sapiens*. Predicting exactly when or how humanity disappears is a fool's errand.

The last extinction event that swept most dinosaurs into oblivion occurred 66 million years ago and was caused by astrological and geological events. The current extinction event is fueled by evolution. Despite obvious and dramatic differences between the two events, their actual time frames may be remarkably similar—around 1 to 2 million years. The last extinction event started out dramatically with the loss of tens and possibly hundreds of thousands of species in a few centuries. As the effects of the asteroid impact and lava flows diminished over the next 2 million years, fewer and fewer species disappeared.

The current extinction event is playing out in exactly the opposite way. It started out with the loss of a few megafaunal species in Africa several million years ago and has gradually picked up steam resulting in more and more extinctions over time. Even though the residual effects of our existence on the planet will result in species continuing to expire in the decades immediately following our extinction, our demise will essentially end the sixth extinction.

The indifference and cruelty humans have inflicted on other organisms sharing the planet with us is going to come back and bite us squarely on the ass. All the laws are currently on the side of economic growth, backed by the will of the people. The upcoming litigation between forests/grasslands/wildlife versus humans will end with humans being convicted of war crimes, and ultimately expelled from the planet forever.

CHAPTER 9

THE PLANET

Most people become personally invested in a small segment of the world. Bird-watching, the stock market, fly-fishing, and human rights are all examples. We all love and cherish the pieces of our individual worlds. The planet is so expansive and beyond our ability to experience it all that few people appreciate that everything we do affects everything else. There is a metaphor for this phenomenon, the butterfly effect. A butterfly beating its wings in Australia, theoretically causes a tornado in North America.

The changes humans have manifested on the planet are so profound that the effects are being felt everywhere. As John Muir noted in 1911, "When we try to pick out anything by itself, we find it hitched to everything else in the universe."

Not only are the effects of the myriad of atrocities humanity is inflicting on the planet being felt everywhere but the whole is greater than the sum of the parts. Much greater. Few see the risk to the planet as a whole rather than to a specific issue. Climatologists study the changing climate. Wildlife biologists research the loss of biodiversity. Toxicologists examine the biological effects of chemicals on living organisms. Epidemiologists investigate infectious diseases. We listen to experts who are themselves focused on a single issue, losing our ability to see that everything is connected.

At the nucleus of all insults to the planet is overpopulation and human behavior. Everything else is electrons revolving around and being driven by the core. The wounds we are inflicting on the planet and ourselves simply because we are human are increasing in intensity and quickly expanding in new and unexpected ways:

- The first issue to raise its ugly head was precipitated by our ancient ancestors: the loss of biodiversity. Increasing evidence suggests it began nearly 2 million years ago but became much more pronounced and identifiable between 35 and 60 thousand years ago when our global population was roughly 1 million people.
- Next on the scene was desertification in the form of concentrated domestic livestock grazing and the excessive use of water for agricultural irrigation. It began approximately 10–15,000 years ago. By then, our population had increased to 10 million people.
- Epidemics date back 5,000 years to two villages in northeastern China: Hamin Mangha and Miaozigou and began when the global population had increased to 27 million. Discoveries suggest that an epidemic ravaged the entire region of northeastern China during that time frame.
- Deforestation also dates back 5,000 years to Iraq in what was one of the first civilizations—Mesopotamia.
- The first pandemic was documented 2,500 years ago when human population had increased to 100 million.
- Climate change began as a result of the Industrial Revolution in about 1800 when the global population had increased to 1 billion people. It was first identified as a problem in the 1930s.
- The use of toxic chemicals became a significant problem 75 years ago when the global population had increased to 2.75 billion people.
- While the production of plastics also began about 1950, microplastics only reached the public's consciousness in 2020.

The increase in the magnitude, number, and frequency of planetary issues mirrors human population growth. Examining each of these topics independently is a distraction and results in consistently underestimating or even ignoring their collective effects. These factors are intertwined and defy an accurate assessment when looked at in isolation.

Climate change and infectious disease are being made worse by deforestation and desertification. Deforestation, desertification, and infectious disease wouldn't be as big a problem if it weren't for climate change. Deforestation can be a precursor to desertification and loss of biodiversity. The factors discussed in this chapter are the proverbial snake swallowing its tail.

The biggest threats to our continued existence are largely amorphous: climate change, biodiversity loss, and toxins in the environment. As a result, they are routinely ignored because they don't seem tangible to most people. However seemingly obscure, they successfully and significantly amplify almost all other threats to our continued existence, including natural events, social unrest, and geopolitical strife.

The reflex, or instinctive behaviors, of *Homo sapiens* is not well-suited to dealing with threats without a clearly defined shape or form. We're evolutionarily hardwired to deal with imminent, concrete, easily visualized threats like predators and wars through struggle or evasion. Once humans do recognize a threat as imminent, our most common reaction is employing a focused approach to deal with the issue to the exclusion of everything else. With numerous factors simultaneously undermining the ability of complex life-forms to exist, this focused response to individual threats is yet another real threat to our continued existence.

There are entire books devoted to climate change, deforestation, over-population, infectious disease, and desertification. Rather than regurgitating what other authors have covered thoroughly, this chapter provides a brief overview of those issues and focuses on often overlooked aspects of the issue. Underappreciated threats that rarely reach the public consciousness are examined in greater detail. The negative effects humanity is inflicting on complex life on the planet described in this chapter aren't exhaustive. Fuel and chemical spills, agriculture, commercial fishing, machine superintelligence, molecular nanotechnology, stratospheric geoengineering, bioengineering, graveyards, renewables, nuclear war and accidents, conventional wars, poaching, and wildlife trade are but a few of the many additional threats not covered in this chapter.

OVERPOPULATION

In 1968, with the release of their book *The Population Bomb*, Paul and Anne Ehrlich were among the first to identify the most significant factor that will precipitate the collapse of humanity. Their book inspired an environmentalist fad in the 1970s. The premise for the book was gradually rationalized away by most as the work of lunatics. It was listed by the *Intercollegiate Review* as one of the 50 worst books of the 20th century. In the *Human Events* list of the "Ten Most Harmful Books of the Nineteenth and Twentieth Centuries," it garnered an 11th-place honorable mention.

Since that time, the global population has more than doubled. During those five decades, humanity has identified six types of quarks, developed the modern internet, eradicated smallpox, decoded the human genome, and developed vaccines for Ebola and COVID-19.

Despite all our new technologies and discoveries, the most basic concept of rapid human growth inside a finite system—our planet—leading to collapse is a concept too difficult for our greatest minds to reconcile. Humanity's carefully calibrated psychological filters go into overdrive to prevent this simple mathematical postulate from entering our psyche. As seemingly prescient as the Ehrlichs were, our collision course with extinction was preordained long before their book was published in 1968.

Money is a figment of the human imagination. Humans attach no value to a habitable Earth but a high value to something that doesn't actually exist. Money is the vehicle we utilize to disengage from the moral and social effects of our decisions. As long as we can frame everything as an economic argument, we don't have to confront the social, moral, or apocalyptic consequences of our decisions—until we do.

Population growth vanished from the agendas of mainstream environmental organizations that previously regarded escalating numbers as a major environmental threat. These groups were primarily shackled by their fear of alienating donors, ultimately selling their purpose and integrity for money. Criticism from progressive and conservative interests claiming that overpopulation is a myth further incentivized these groups to pretend the rising global population wasn't a factor in planetary degradation.

The fear of losing money disabled those organizations best placed to understand the ultimate consequences of thinking only about money. Even at environmental meetings and conferences where attendees are almost exclusively "pro-environment," those with the courage to broach the topic are usually the recipient of verbal attacks and abuse.

Those who defend the belief that overpopulation isn't at the core of every environmental problem are not unique. Every myth presents itself as an authoritative, factual account, no matter how much the topic varies from natural law or ordinary experience. There is a long, bloody history of *Homo sapiens* defending myths against those who might either question their veracity or have a competing myth. It has resulted in the deaths of millions of people since the dawn of the Agricultural Revolution.

As a result of ignoring the obvious reality, newborns are effectively positioned as moneymaking machines. The former prime minister of Japan suggested that women who bore no children should be barred from receiving pensions. In most countries, those who choose not to have children are required to pay for those who do through taxes. In this campaign for more babies, childbearing is reduced to a means for economic growth.

Even though overpopulation, natural resource extraction, and environmental degradation are clearly linked, the needs of the economic market trump the needs of the planet. Children are nothing more than moneymakers in the eyes of politicians, forever blind to the moral, environmental, or humanitarian consequences of their policies. Market thinking has obliterated moral thinking on a grand scale. After all, if the West doesn't produce more children, it can't produce the wealth needed to look after parents when they retire.

No social animal is ever guided by the interests of the entire species to which it belongs. No pika cares about the interests of the pika species; no northern spotted owl will lift a feather for the global northern spotted owl community; no wolf alpha male makes a bid for becoming the king of all wolves. Likewise, few humans care about the interests of *Homo sapiens*. People only care about themselves and those who directly affect their lives.

As the Ehrlichs discovered, even the most benign suggestion of slowing population growth is met with derision, criticism, ridicule, and mockery. In what was either one of the most courageous statements ever made by an American politician or a naive utterance of the obvious, Representative Alexandria Ocasio-Cortez of New York suggested it may not be ethical to have children. She asked the question in an Instagram post as to whether we should have children, knowing their lives are going to be problematic.

Given the caustic feedback received by those broaching the topic at environmental conferences, the righteous indignation and vitriol in the responses to the congresswoman's statement from the public on social media were entirely predictable. She was accused of being fascistic, admitting to civilizational suicide, and not believing in her country, values, or free-market economics. The only issues left out of the cyberattacks were she didn't believe in mother and apple pie, and those were probably oversights.

The comment to her Instagram post that best reflects human reaction to the problem was this one: "I don't read the science updates anymore because it is too awful. I just don't engage with that because it is too hard to reconcile with my choices." The commenter's response represents the attitude of the majority of humanity: see no evil, hear no evil.

Interestingly, someone who saves another person from death is considered a hero, but anyone who tries to save humanity from extinction is vilified, denounced, and censured. Acknowledging a problem of the magnitude of human extinction threatens the foundation of the carefully constructed myths we use to define our lives. For many people, even raising the issue threatens every fiber of their being.

Like the congresswoman, there are many concerned about the already crippling environmental burdens placed on future generations. Unfortunately, those worries would have been most appropriate centuries ago. Humanity has moved past the point where future generations are an issue. If there are any, they will be few in number.

Ignoring the inevitable is something all species do. Humans are no different than bacteria in a Petri dish or a mountain pine beetle in a climate-stressed lodgepole pine forest. All available resources are utilized to grow and reproduce until the inevitable collapse. We're traveling down the same path as all species that have gone through exponential increases. The only difference is *Homo sapiens* is doing it on a grander scale. The Herculean ability to ignore the greatest threat to our existence would be comical if not for the rapidly approaching consequential conclusion to our existence.

TRASH, WASTE, AND TOXINS

The single best measure of a person's commitment to the environment isn't what they eat or drive, their political affiliation, or whether they say or perceive themselves as an environmentalist. It's how much waste, emissions, and toxins they personally cause. The more trash, emissions, and toxic waste someone generates, the more negative effects that person has on the planet. It is a direct reflection of how much mining, oil and gas development, logging, land degradation, and manufacturing is necessary to support a person's lifestyle.

Modern society is poised to drown either figuratively or literally in our own trash, emissions, and toxic waste. While there are many who view climate change as the most serious potential existential crisis, trash, waste, and toxins are even more dangerous. Their collective potential to create an unmitigable crisis may take a little longer to manifest itself than climate change, but are more of a certainty. The more urbanized and industrialized a country becomes, the more trash it produces. Trash and waste aren't just litter—they are also a slow, toxic poison.

Some of the leading international economists look at the amount of garbage transport as a direct indicator of how well the economy is doing. The more economic growth, the more trash we engender. There is absolutely nothing we do that doesn't generate emissions, trash, and waste.

Economic growth requires energy and raw materials. Capitalists correctly point out that while humankind's use of energy and raw materials has mushroomed in the last few centuries, the amounts available for exploitation have increased. Instead of relying exclusively on iron and wood for materials, we now use plastic, rubber, aluminum, silicon, and titanium. While we rely heavily on fossil fuels, nuclear power, and hydroelectric energy, we're supplementing those with wind, geothermal, and solar energy.

However ingenious we've been with acquiring new kinds of energy and raw materials, humans have continually ignored or swept the waste and emissions from all energy production and manufacturing under the carpet. The carpet is wearing out in a myriad of ways. Global warming, for example, is largely the result of wastes associated with burning fossil fuels and would be just as appropriate in this section as having its own.

The average person on the planet generates 555 pounds (252 kilograms) of municipal solid waste annually with more than a third of that not managed in an environmentally safe manner. Worldwide municipal solid waste is expected to grow from the current 2 billion metric tons to 3.4 billion tons by 2050. Developed countries produce more waste per capita because they have higher levels of consumption.

There may no longer be a home anywhere on the planet not containing toxic chemicals linked to allergies, birth defects, cancer, neurotoxicity, hormone disruption, and psychological abnormalities. Asbestos, benzoyl benzoate, lead, formaldehyde, and fire retardants in foam furniture, carpet padding, electronics, plastics, textiles, and building materials are just a few. Whenever there is a major flood anywhere in the world, photos of inundated homes are shown on the news. What isn't shown is the toxic stew of chemicals stored or embedded in each home that entered the water that will eventually return to the river and ocean or seep into groundwater.

Fish populations throughout the world's oceans are contaminated with industrial and agricultural pollutants, collectively known as persistent organic pollutants (POPs) and include dioxins and PCBs. They can be anywhere and in any species of marine fish. There is a high degree of variability in concentrations of POPs between regions and groups of fish.

Methyl mercury can impair cognitive thinking, memory, attention, language, and fine-motor and visual-spatial skills in humans. Algae absorb organic methylmercury, so the fish that eat algae will also absorb this toxic substance. When larger fish at the top of the food chain eat these fish, they too accumulate methylmercury. As ocean temperatures warm, fish expend more energy swimming requiring eating a greater number of smaller fish to meet metabolic needs. The result is fish having an ever-increasing intake and accumulation of methylmercury. For example, between 2012 and 2017, researchers found that Atlantic bluefin tuna saw an increase of methylmercury levels by as much as 3.5 percent each year. The Atlantic, Indian, and Northern Pacific oceans are the fastest-warming oceans on the planet, but the exposure increases are happening everywhere.

Exposure to chemicals found in trash, especially plastics, have been linked to issues in reproductive health, obesity, increased allergies, diabetes, and cancer. Unlike traditional materials, plastic won't dissolve and rust or break down in any useful time frame. The plastic we've produced will still be here hundreds and thousands of years from now in litter, layers of landfill, and detritus in the oceans. When not disposed of properly, it gradually breaks down into smaller and smaller pieces. The marine environment is especially at risk from decomposing plastics, with many animal species swallowing microscopic bits. When marine fish and animals aren't eating plastic, they are becoming entangled in it.

Humans have produced more than 8 billion tons of plastic, mostly since the 1950s. Less than 10 percent of it has been recycled. Over time, much of it has broken down into tiny plastic particles, called microplastics and nanoplastics, that eventually contaminate our food and water. Researchers find microplastics everywhere they look: in the air, soil, rivers, deepest oceans, and remote corners of the Arctic.

Every year, 8 million tons of garbage is added to our oceans. There are over 5 trillion pieces of plastic debris in the ocean. Of that mass, 269,000 tons float on the surface. A new study by the Pew Charitable Trust predicts the annual flow of plastic into the world's oceans will nearly triple over the next 20 years.

Nearly 10 percent of microplastics dispersed in the ocean each year come from textiles. Garments are a huge source of microplastics because so many are now made of nylon or polyester. Each wash-and-dry cycle, these garments shed micro filaments that move through our sewage systems and end up in waterways. It is estimated that half a million tons of these contaminants reach the ocean each year.

When anyone eats a bite of food or has a drink of water, they're almost certainly taking in these tiny plastic particles. Microplastics are so prevalent that we breathe in tens of thousands of tiny plastic fragments or fibers every year. Researchers at the University of Newcastle in Australia estimated that the average person now absorbs through food, water, and air about five grams of plastic a week. This is equivalent to ingesting a credit card a week.

There is increasing evidence that pollutants piggyback on microplastics, either by being absorbed or attaching to their surface. As vectors for other toxic substances, microplastics in some geographic areas may be extremely dangerous to humans, plants, and animals.

Because ingestion of microplastics is a relatively recent phenomenon, scientific evidence of the effects on humans is in its infancy. Already there is research pointing to the chemical components of microplastics being endocrine disrupters. The question isn't whether it is hazardous to our health, but at what level it causes serious issues. Those problems could manifest themselves in a variety of ways and will vary from individual to individual.

Just like climate change and many other threats to the planet, even if we stopped all use of plastics today, the continued decomposition of those already in the environment will increase microplastics in the air and water for decades. The amount of plastic we generate isn't going to stop or even slow down—it will continue to increase. It's the cost of economic prosperity.

By the time the research is available to understand the health effects from a person ingesting a credit card's worth of plastic each week, we'll probably be ingesting two or three. How much plastic can we ingest and still survive? We have become the subject of our own ongoing clinical trial.

Most radioactive waste is generated in countries with nuclear power plants, nuclear armament, or nuclear fuel treatment plants. High-level waste is produced by nuclear reactors and accounts for over 95 percent of the total radioactivity produced in the process of nuclear electricity generation. Spent fuel is highly radioactive and often hot. The amount of high-level waste worldwide is currently increasing by about 12,000 metric tons every year. A 1,000-megawatt nuclear power plant, which can satisfy the electrical needs of a city the size of Florence, Italy, or Gaza, Palestine, produces about 27 tons of spent nuclear fuel annually.

Most scientists agree that the main proposed long-term solution to radioactive waste is deep geological burial. However, as of 2022, Finland is the only country in the advanced stage of the construction of such a facility. Proper disposal of nuclear waste and reducing emissions generated from burning fossil fuels will still be on humanity's to-do list as we slide into an ecological Armageddon.

Of the 100 billion garments produced each year on Earth, 92 million metric tons end up in landfills. Textile waste is expected to balloon to 134 million metric tons by the end of the decade. The apparel industry's global emissions will increase by 50 percent by 2030. The number of times a garment is worn has declined by around 36 percent in 15 years. It takes 20,000 liters of water to produce one kilogram of cotton.

Furniture, carpet, and rug waste is estimated at 75 million metric tons annually. Only about half of textile and furniture waste end up in landfills.

The remainder is dumped wherever humans find it convenient to do so, harming the environment in the process.

Mexico generates more trash than any other Latin American country. It is estimated that 70 percent of all garbage in the country ends up in illegal dumps. Ravines, forests, riverbanks, the side of highways, and vacant lots all serve as garbage dumps creating environmental damage and threatening public health. Mexico isn't alone. Four billion people worldwide, and 40 percent of the planet's trash and waste is discarded anywhere convenient.

There is discarded trash in the wildest places left on the planet. Thoughtless fire crews, timber crews, oil and gas developers, and recreationists all contribute to defiling remote landscapes with cans, bottles, wrappers, and other garbage.

Hazardous chemicals are now found in the tissue of every person on the planet. Humans emit more than 2.5 billion metric tons of chemical substances a year, in a toxic avalanche that is harming people and life everywhere on the planet. There is more than 220 billion metric tons of chemical waste from all sources including eroded soil, mineral waste, and construction spoil. Toxic waste is a dangerous byproduct of manufacturing, farming, water treatment systems, construction, automotive garages, laboratories, hospitals, and other industries. Households also generate hazardous waste including pesticides, herbicides, leftover paints, batteries, cosmetics, and used electronic equipment. These toxins can harm people, animals, and plants in a variety of ways. Toxins such as mercury and lead accumulate in the environment over time and persist for long periods, bioaccumulating in the tissues of fish and other animals.

WHO estimates that 13.7 million people die annually from "environmental chemicals." Recently, there has been a shift to move chemical production to poor countries that have even fewer environmental regulations and lack the resources to effectively deal with toxins in the environment.

While people focus on climate change or biodiversity loss or deforestation without considering their effects in combination, we do the same with toxins. The negative effects of mercury or DDT or dioxins are considered independently while completely ignoring that chemicals act in combination, occur in mixtures, and undergo constant change. A chemical that is not toxic and is combined with several other chemicals may contribute a much larger risk to the health and safety of the whole population and environment.

Roughly 2,000 new chemicals are introduced each year. The effects of these chemicals to the planet, humanity, fish, wildlife, and other organisms are unknown. Even if a great deal of research eventually does determine their safety, it is impossible to know if the effects of these chemicals in combination with the thousands of chemicals already in existence is harmful.

Despite attempts to regulate chemical use, only 21 out of 350,000 chemicals have so far been banned. There are now attempts being made to roll back the anemic chemical regulations already in place, further exposing people and the planet to ever-greater risks. Humanity's lasting legacy to the planet won't be our cherished creations but the mountains of garbage and the toxic stew we leave behind.

CLIMATE CHANGE

There is abundant scientific evidence that human activity is the primary contributor to rising global temperatures. Increased temperatures at Earth's surface, in the atmosphere, and in oceans are well-documented. The planet's climate is now changing faster than at any point in the history of modern civilization and has already resulted in a wide range of impacts across every continent.

The amount of human-caused greenhouse gases continues to increase simultaneously with a decline in the planet's ability to pull and store carbon dioxide from the atmosphere. As the chasm widens between these two factors, there is an increasing cascade of negative effects to complex organisms inhabiting the planet.

In the 1930s, British engineer Guy Stewart Callendar noted that the United States and the North Atlantic region had warmed significantly during the Industrial Revolution. He continued to argue into the 1960s that a greenhouse-effect warming of the planet was underway. In the 1970s, scientists theorized pollution could block sunlight and cool Earth. In fact, Earth did cool somewhat between 1940 and 1970 due to a postwar boom in aerosol pollutants, which reflected sunlight away from the planet. As the brief cooling period ended, temperatures resumed their upward climb. The early 1980s saw a sharp increase in global temperatures. In 1987, the Black Dragon Fire in China was the largest wildfire ever recorded at that time—around 27,800 square miles (72,000 square kilometers). Since that time, many years have now exceeded the temperatures documented in the '80s. In 2019 and 2020, the Australian Bushfires and Siberian Taiga wildfire burned 79,150 square miles (205,000 square kilometers).

Aerosol pollutants are only a problem for those who enjoy breathing air. At high concentrations, these tiny particles are one of the deadliest substances in existence, burrowing deep into our bodies where they damage heart and lungs. Air pollution from burning coal, driving cars, and using fire to clear land, among other activities, is the fourth leading cause of death worldwide, killing about 5.5 million people each year. It is estimated that the cooling effect of aerosols is so large that it has masked as much as half of the warming effect from greenhouse gases. The choice between climate change and aerosol pollutants is a devil's bargain.

The first global agreement to reduce greenhouse gases, the Kyoto Protocol was adopted in 1997. Beginning with the Bush administration in 2000, climate change in the United States has seen two decades of politicization, foot-dragging, and denial. A similar response has occurred in many other countries around the world. Today, atmospheric carbon dioxide levels are higher than at any point in at least the past 3 million years.

Emissions of heat-trapping greenhouse gases from fossil fuel combustion, deforestation, and land-use change are the primary drivers of global warming. Human health, agriculture and food security, water supply, transportation, and energy are all expected to be disrupted at increasing levels.

Forests, barrier beaches, and wetland ecosystems can buffer the impacts of extreme events like fires, floods, and severe storms. These ecosystems are so compromised by human activities over the last 300 years they have lost their ability to deal with the effects of climate change. Only tiny remnants of the planet's ancient forests remain. It is estimated that 1,312,750 square miles (3.4 million square kilometers) of inland wetlands have been lost since 1700, primarily from conversion to croplands, although the losses from urbanization and infrastructure are also significant. Extreme weather events and wildfires; decreased air quality; and diseases transmitted by animals, food, and water are becoming more pronounced and will continually worsen.

Over the last two decades, climate scientists have consistently underestimated the speed and magnitude of global warming. The global climate is a complex system that responds to many inputs, not all of which are well understood. The numerous variables affecting global warming usually integrate with each other to amplify the change over what could be expected by research looking at each of them independently. Climate change is influenced by almost all human activities.

When the implications of scientific findings are significant and far-reaching, they come under extraordinary scrutiny. Research funding is often premised on the rigor of their findings. Scientists, out of necessity, are overly cautious and as a result routinely underestimate the climate response.

A part of the reason scientists are overly cautious is because industries adversely affected by research findings routinely hire other scientists to undermine the credibility and results of legitimate research. These scientists for hire are endearingly known in the biological community as biostitutes.

Candid disagreement has always been an important vehicle for furthering scientific progress. However, an agenda for disagreement fueled by profit has the opposite effect, stifling the sharing of ideas and information and forcing scientists to chronically underestimate effects.

There is also an asymmetry in how scientists think about error and its effects on their reputations. Many scientists worry if they overestimate a threat, they will lose credibility. The flip side is if they underestimate, it will have little reputational impact.

Finally, there is a perceived need for consensus among the scientific community. Many scientists worry that if disagreement is publicly aired, government officials will conflate differences of opinion with ignorance and use this as justification for inaction. Others worry that even if policy makers want to act, they will find it difficult to do so if scientists fail to send an unambiguous message. Both concerns are legitimate and play out regularly in public discourse.

Even if zero emissions occurred starting tomorrow, ocean and air temperatures will continue to rise because the greenhouse gases we've poured into the atmosphere for decades won't disappear overnight. A percentage of the emissions will still be warming the planet 500 years from now. Climate change is resulting in extreme rainfall in some regions on the planet and more severe drought in others. It is making it more difficult for water managers to ensure we have enough to drink and a reliable supply of water for farmers to produce the food we eat. Longer and hotter heat waves and freezing polar vortexes will continue to push electrical grids to and beyond their limit.

Plants typically absorb carbon dioxide and release oxygen, but they also "exhale" carbon dioxide, particularly at night when photosynthesis ceases due to the lack of sunlight. Increasing temperatures stress forests causing them to emit more carbon dioxide. Continued warming will reduce photosynthesis and exponentially increase the amount of carbon dioxide trees release through respiration as they try to breathe enough to deal with insects, disease, drought, and warmer temperatures.

It isn't necessary to be a mathematician at a university to know that there is no chance of humanity reducing carbon dioxide emissions in a timely manner. All it takes is a mastery of fourth-grade math. To keep the possibility of limiting global warming levels to 1.5 degrees centigrade, global greenhouse gas emissions need to drop 55 percent by 2030. Between

2016 and 2018, there were no increases or reductions, but in 2019 they rose by a little over 1 percent.

Reduced travel because of the coronavirus pandemic resulted in a 7 percent reduction in global carbon emissions in 2020. Despite the reduction in emissions, planetary carbon dioxide levels in the atmosphere continued to increase as a result of logging, wildfire, mining, damming rivers, and the conversion of forests for development, agriculture, and livestock production.

Ignoring the combined effect of increased emissions and the increasingly compromised capability of the planet to absorb carbon dioxide has humanity firmly ensconced on the train to nowhere. Currently, total emissions hover around 53 gigatons of equivalent carbon dioxide annually. To meet the goal, humans need to reduce emissions to 29 gigatons.

Changes in Earth's orbit have operated slowly over millions of years modifying the amount of solar radiation reaching each latitudinal band of Earth during each month. The glacial-interglacial cycles Earth has experienced for the past 2.6 million years occur on 100,000-year intervals. History has shown that during each interglacial period, carbon dioxide levels increase and global temperatures warm. Carbon dioxide levels typically peak at between 275 and 300 parts per million (ppm) before temperatures once again cool off and carbon dioxide levels drop as the planet enters another glacial period. This natural, proven, and predictable warming mechanism cannot be responsible for the current atmospheric carbon dioxide level of 425 ppm or the sudden spike in global temperatures.

A Canadian study found that having a child generates an estimated 58 tons of carbon dioxide emissions every year. In contrast, going car-free saves a minuscule-by-comparison 2.5 tons. The planet's population will increase by approximately 75 million people next year, resulting in nearly five gigatons of carbon dioxide being added into the atmosphere.

There are millions more vehicles on the planet every year. Between the additional population growth and increased number of vehicles, in a single year, humans are adding over five gigatons of carbon into the atmosphere. Is it any wonder we can't and won't make progress toward carbon emission reduction goals?

Until recently, it was thought that the entire Amazon rainforest absorbed about two gigatons of carbon dioxide from the atmosphere each year. Now, even that is in question. The Amazon, suffering from land-clearing fires and drying wetlands and soil compaction from logging, may now be worsening climate change as a result in increases in the release of the greenhouse gases methane and nitrous oxide. While the expansive Amazon rainforest is still one of the world's largest carbon sinks, its ability to scrub human-caused greenhouse gases from the atmosphere is diminishing rapidly.

On a per-acre basis, the Amazon was not nearly as efficient at absorbing carbon as the coastal temperate rainforests of North America. The vanishing Douglas fir forests of Oregon, Washington, and British Columbia and the hemlock and cedar forests of Alaska and British Columbia store about twice as much carbon per acre as the Amazon. The giant redwoods of Northern California, which store seven times as much, are regarded as the most carbon-dense forests in the world.

If all the ancient forests in the world were intact, each continent would have the capability to remove between four and six gigatons of carbon dioxide from the atmosphere. Intact grasslands like the tallgrass prairie in the middle of North America might have at one time added an additional gigaton of carbon sequestration buffer in each continent.

Logging, grazing, and the conversion of wildland areas to cities, developments, and highways have resulted in the planet losing about two-thirds of its natural land-based capacity to remove carbon dioxide from the atmosphere. The amount of carbon absorbed by forests everywhere continues to decline as the remaining minuscule tracts of ancient forests are logged and wildfires increase. Instead of having the capacity to remove approximately 40 gigatons of carbon dioxide annually, land-based areas on the planet can now only remove 12 to 15 gigatons. Estimates are oceans remove another 15 to 17 gigatons. While we're emitting 53 gigatons of carbon dioxide into the atmosphere, only about 30 are being removed. There is an ever-widening gap between the planet's finite capacity for carbon storage and photosynthesis and the emissions humans release into the atmosphere. When it comes to carbon dioxide in the air, humanity is in a deficit spending mode that would make the most fiscally irresponsible governments seem frugal by comparison. Humanity has no mechanism to reverse course. People will continue to increase their contributions to climate change while sucking on the pacifier of economic growth.

BIODIVERSITY LOSS

Biodiversity, or biological diversity, is the variety and variability of life on Earth. Biodiversity is a measure of variation at the genetic, species, and habitat or ecosystem level. Much of Earth's great biodiversity has already disappeared. Many species have been lost without humanity even knowing they existed.

It is no longer possible to convince most politicians or ordinary citizens of the importance of conserving biodiversity. In today's urban jungle, these arguments make no real connection to people's own experience. Knowing there are at least a few million species, many shrug at a few thousand species going extinct every year. Will it have any influence on the halftime show at the Super Bowl this year? Probably not.

Many people are "animal lovers" and feverishly attached to their pets, but hearing about the thousands of species of insects inhabiting decaying tree trunks causes their eyes to glaze over. There are a few people who appreciate the intrinsic value of species and biodiversity. They, like me, are frightened by how thoughtlessly and uncaringly we are steamrolling millions of years of evolutionary biodiversity. The greater the loss of biodiversity, the less the resilience and adaptability of the planet's ecosystems to the changes wrought by humanity.

There are many individual, but interacting, factors contributing to biodiversity loss. The overall driver is what ecologists term competitive displacement of nonhuman life by the inexorable growth of the human enterprise—in other words, overpopulation. Habitat loss, invasive species, climate change, deforestation, desertification, urbanization, and pollution collectively play a supporting role. While animals like tigers, lions, and elephants get more attention, the collapse of humanity will likely be much more closely tied to the loss of forests, plants, insects, bacteria, and fungi.

Photosynthesis is a process used by plants, algae, and photosynthetic bacteria to convert light energy into chemical energy that can later be released to fuel their activities. On a finite planet where a great number of organisms share the same space and depend on photosynthesis either directly or indirectly, the continuous expansion of one species always results in the contraction of others. There is always a negative effect on the planet whenever there is human population growth. As can be witnessed by the reverse engineering example of the seemingly benign bowl of rice, more people need more food, clothes, shelter, medicine, and water, which creates an ever-greater strain on the environment in a myriad of ways.

Biological processes are responsible for supporting the life-friendly chemical balance of the oceans. Photosynthetic bacteria and green plants have stocked and maintained Earth's atmosphere with the oxygen necessary for the evolution of animals. Photosynthesis gradually extracted billions of tons of carbon from the atmosphere so that Earth's average temperature has remained for geologic ages in the narrow range that makes water-based life possible. Countless species of bacteria and fungi continuously regenerate the soils that grow our food. Life runs right alongside us, unseen.

Those viruses, bacteria, fungi, and parasites that cause disease are nearly everywhere. Healthy wildlife and ecosystems have evolved defenses to fend off most diseases before they have devastating impacts. Biodiversity ensures both greater resilience and resistance to the impacts of disease because there are more possibilities for defying and surviving the outbreak. Biodiversity is becoming more important at the same moment in time it is deteriorating rapidly. Ecosystems are not only dealing with natural environmental change but are being bombarded with deforestation, desertification, invasive species, and climate change.

Insects are in trouble. Maybe the single most alarming, documented biodiversity issue is the reported decline in insect populations. In a 30-year time frame, flying insect populations in Germany have declined by 75 percent within 63 protected areas. The same trend appears in other parts of Europe and Canada where data is available. Half of all moth and butterfly species are in decline and one-third are threatened with extinction. Relatively pristine rainforests in Puerto Rico showed a similar decline. Because they lay their eggs in water, aquatic insects are even more vulnerable to pollution and development than their terrestrial counterparts.

Why should we be surprised? Earth has lost 60 percent of all native animals since 1970. Insect biodiversity and biomass are declining at approximately the same rate as native mammals, birds, fish, amphibians, and reptiles. They all rely on the same thing we do: clean water and air, and native plants and trees.

Insects are an important food resource for many birds, small mammals, and fish. Over 75 percent of flowering plants are pollinated by insects, birds, and bats. Thirty-five percent of the world's crop production is more or less dependent on bird, bat, and insect pollination. People depend on a rich diversity of life-forms to provide various life-support functions essential to their existence and continued survival.

Although resistance to pesticides is quick to evolve among the pests, it is less likely to evolve in species we don't target. This results in two problems. The first is the simple loss of the biodiversity around us on which wild ecosystems depend. The second is among the species killed by pesticides are those that benefit us, including natural enemies of the pests we are trying to control. Without native bees, for example, we would face a diet lacking in vital nutrients, entire food chains depending on animals that eat seeds and berries would disappear, colorful flowering plants would vanish, and ecosystems would collapse.

The importance of native birds and other animals will be appreciated by humanity only during and after their loss. For example, after the extinction of the dodo and Mauritius giant tortoises on the small volcanic island of Mauritius, the tambalacoque tree, whose fruits had been a primary food item, also declined precipitously. The digestive process of the now-extinct animals may have helped distribute and germinate the seeds. Even though a limited number of other animals were capable of the ecosystem function the extinct animals provided, they were too few in number to effectively fill the void and prevent declines in tree numbers in some areas where they were historically prevalent. The extinction of the thylacine in Australia and giant ground sloth in South America had even more wide-ranging and profound effects on other plants, animals, and ecosystems.

We are trading a richness of butterflies, bees, ants, moths, and other wild species for chemical-resistant cockroaches, bedbugs, lice, houseflies, and fleas. We're replacing superb fairy-wrens, red-crested cardinals, tree swallows, bluebirds, and purple martins with starlings. Domestic and feral cats replace owls, hawks, badgers, foxes, weasels, potoroos, quokkas, chinchillas, gerbils, golden moles, mice, shrews, and voles.

Many natural areas where species have persisted for thousands of years have lost their ability to provide essential food, water, and security that allowed them to survive and flourish. An increasing number of animals like the Attwater's prairie chicken, northern white rhinoceros, and Rio Grande silvery minnow are imprisoned, maintained only by the constant intervention of human keepers.

As humans continue to destroy and simplify habitat through deforestation, urbanization, desertification, and the use of chemicals to control and kill pests, the life-forms that support the human enterprise will continue to decline and wink out of existence.

DEFORESTATION

Deforestation refers to the decrease in forest areas across the world that are lost for other uses such as agricultural croplands, urbanization, and grazing. Ninety-five percent of global deforestation occurs in the tropics. Logging in theory, doesn't decrease forest areas, because seedlings are re-planted. However, logging not only results in the loss of carbon storage, but also in carbon dioxide emissions from the manufacture and operation of industrial machinery. Logging and deforestation are both responsible for significant reductions in biodiversity and bioabundance. In fact, because of climate change and lack of soil productivity following logging, the impacts of logging and deforestation are nearly identical. Still, perception gives Canada, Europe, Russia, and the United States an "out" since during the last century they have done very little deforestation. They can sit back and criticize countries in the tropics while busily logging their own forests. A natural, old-growth forest is an order of magnitude different from one that has been razed by logging.

The immediate results of deforestation and logging include major changes to the Earth's landscape and diminishing biodiversity and wildlife habitat. Carbon sequestration is the long-term process involved in carbon capture and storage of atmospheric carbon dioxide and other forms of carbon that contribute to climate change. Carbon is stored in places like oceans, vegetation, and soils which are commonly referred to as carbon sinks. Forests absorb carbon dioxide from the atmosphere and store it in tree trunks, roots, and soil.

Logging results in a reduction of carbon sequestration, which in turn contributes to climate change. Deforestation and forest damage from log-ging currently removes about a gigaton of carbon sequestration capacity from the planet every year. In some regions, deforestation and logging are currently the biggest contributors to global emissions, slightly ahead of road transport and fuel and power for residential buildings.

Although young forests sequester carbon at a faster rate, the volume they sequester is relatively small and insufficient to compensate for carbon losses through soil respiration, typically resulting in a net release in carbon. It is not how fast carbon deposits are made but how big the deposits. The volume of carbon sequestration of older forests is much greater than young ones.

Managed forests, places where logging has taken place and trees have been planted, capture and hold much less carbon than unmanaged forests. Old trees not only store more carbon and prevent it from escaping into the atmosphere but actively convert carbon dioxide in the air into their trunks, branches, roots, and leaves.

As a result of the disturbance and disruption of the forest ecosystem, the huge amounts of carbon older forests store are quickly returned to the atmosphere when logged and planted with young trees. Logging releases nearly half the carbon and puts it into the atmosphere within years. When trees are cut down, they immediately start to decay, and a significant amount of the carbon stored in their above-ground biomass is released into the atmosphere. The carbon in the soils, stumps, and roots is released over time, facilitated by microorganisms and other soil organisms that break down the organic matter.

Another 25 percent of a logged tree's carbon is lost during manufacturing. As the finished wood products decay over time, they emit even more. Buildings outlive their usefulness and may be replaced in 30–50 years. When buildings are replaced or demolished, the carbon stored within them is released back into the atmosphere, contributing further to carbon emissions. Old-growth forests accumulate and hold large quantities of carbon for centuries. Once disturbed, much of this carbon, even soil carbon, moves back to the atmosphere.

Tree seedlings are planted to try to reverse the loss of carbon storage, presuming that drought, pests, disease, and wildfires associated with a rapidly warming climate don't undo those efforts. Forests around the world are losing resilience to climate warming.

A 2017 study found that the maximum potential of natural climate solutions would be almost 24 gigatons of carbon per year while simultaneously safeguarding food security and biodiversity. The cheapest natural climate solution is protecting old-growth forests. The timber industry is no different than the fossil fuel industry in its effective and concerted efforts to hide the true financial and ecological costs of logging.

In the state of Oregon, research has shown that greenhouse gas emissions from timber harvest and logging are 50 percent higher than those generated from all other forms of transportation. Chainsaws, logging trucks, lathes, and other machinery used for logging contribute 33 million tons of carbon dioxide to the air. This figure doesn't include the fossil fuels necessary for the mining and manufacturing to create the machinery. As a comparison, the dirtiest coal plant in the world, Taiwan's 5,500-megawatt Taichung Power Plant, spews about 36 million tons per year. Logging is thus a significant contributor to climate change in a myriad of ways.

Logging has degraded many of the world's forests and life within them in a relatively short period of time. While some untouched forests remain, the timber industry lustfully eyes these remaining tracts as a source of easy, short-term profit. Between 2018 and 2021, the planet lost ancient forests in an area the size of Belgium annually.

Barely 20 percent of old-growth forests remain, and the South American rainforests account for a significant portion of these. Approximately 17–20 percent of the Amazon rainforest has been deforested since the 1970s. People from all parts of the world, particularly the United States, Canada, and Europe are blind to their hypocrisy for insisting South American countries conserve their forests while busily annihilating our own.

In the last 150 years, deforestation, logging, and the burning of fossil fuels have raised the amount of carbon dioxide in the atmosphere by more than 25 percent. The pace of logging is dictated by consumer demand for lumber and wood products. Forests ultimately bear the burden of the increasing consumption of lumber.

The logging industry, politicians, the real estate industry, and local communities that have been supported by logging see forests as a source of wealth, economic growth, and jobs. Those people viewing forests as a source of recreation and living systems that maintain the planet's capacity for supporting complex life are increasingly in the minority. The inability to monetize their argument along with a global population increasingly separated from their habitat continually increases the divide between the two groups. As the global population continues to climb and the demand for more lumber increases, the winners and losers in the conflict are easy to predict.

Logging roads speed up soil erosion, wash away fertile topsoil, pollute streams and rivers, and increase the likelihood of human-caused wildfires. Illegal logging has become a problem with the growing worldwide demand for lumber. Estimates indicate that up to 70 percent of the wood harvested in Indonesia and 25 percent from Russia comes from illegal logging practices.

Between 1958 and 1998, China virtually liquidated its forests to facilitate industrial and agricultural expansion. Devastating floods in 1998 that killed over 3,700 people and left 18 million homeless illuminated the consequences of deforestation of large areas. The last two decades has seen China taking steps to reforest areas and minimize illegal logging.

INFECTIOUS DISEASES

Microbes that cause diseases can't survive long among sparse populations. It wasn't until the global population had increased to 100 million people that it became possible for humanity to graduate from epidemics to pandemics. Unlike epidemics, pandemics require a contagion to be spread over multiple countries or continents rather than a community, population, or region. Historically, children were most often the victim of such diseases. Our current population, density, and mobility have resulted in a more recent transition from pandemics into a new phase of perpetual pandemics. There have been three pandemics so far this century, COVID-19, H1N1 influenza, and SARS.

The role of infectious agents was discovered in the late 19th century, but humans have long known that climatic conditions affect epidemic diseases. Roman aristocrats retreated to hill resorts each summer to avoid malaria. South Asians used strongly curried foods in high summer to combat diarrhea. The plagues of Antonines, Cyprian, and Justinian were all pandemics, probably originating in China/Mongolia and transferred via the Silk Road. Pandemic diseases may have taken out more than 90 percent of the population in the Americas in the 16th century.

Climate change, the loss of biodiversity from mass removal of trees and entire ecosystems, the fragmentation of habitats, and population density (an effect of overpopulation) are influencing the emergence, reemergence, and expansion of infectious diseases. Climate change has expanded the amount of area with optimal temperatures for ticks and mosquitoes to spread vector-borne diseases including malaria, West Nile virus, lyme disease, and dengue fever, to name just a few.

Climate change and the loss of biodiversity increases the probability of virus spillover from wildlife to other wildlife species and to humans. Recent examples are zoonotic diseases like Ebola and COVID-19. Industrial expansion along with climate change displaces more mobile wildlife into shifting from their natural habitat and readapting in areas where they come into contact with species never before encountered. This creates opportunities for viral sharing between previously isolated species and humans.

While the inability to move into higher elevations to escape warmer temperatures will inevitably lead to the extinction of the American pika, many other mammals existing in lower elevations will move into those high-elevation habitats to avoid the heat. Because of their limited geographic extent globally, high-elevation habitats will become points of congregation and biodiversity hot spots leading to a higher risk of novel virus sharing between mammals. Whether climate change forces mammals to shift latitudinally or into ever higher elevations, the rapid forced movements into new species assemblages will increase the chances of viral spillover between themselves and ultimately into humans.

A recent analysis by Chinese researchers found that the emergence of new pathogens tended to happen in places where a dense human population had been changing the landscape by building roads and mines, cutting down forests, and intensifying agriculture. China, India, Nigeria, and Brazil are all hot spots, but this description is a scenario playing out on every continent on the planet with the continual expansion of urban areas. It is synonymous with all aspects of economic growth.

Humans modify and disrupt the ecosystems, sometimes killing native animals and other times caging them. They are sent to market either for food or as part of legal or illegal wildlife trade. Viruses quickly jump from their natural hosts to whatever is at hand. Because of our ubiquity, convenience, and abundance, people now provide viruses with one of their best opportunities. There are as many as 600,000 species of mammal viruses circulating in wildlife that could potentially spread to humans but that are largely undetected.

The Centers for Disease Control and Prevention (CDC) estimates that three-quarters of new or emerging diseases that infect humans originate in animals. We destroy natural ecosystems at our own risk. Infectious disease transference is but one of many. However bad we think coronavirus is now, viral pathogens will continue to exact an even heavier toll as climate change, deforestation, and biodiversity loss worsens.

Air pollution, aeroallergens, drought, famine, lack of clean water and food, compromised health infrastructure due to natural or human-caused factors, toxins in the environment, and natural disasters can exacerbate an infectious outbreak or pandemic. Many of these factors are already compromising human immune function, making people more susceptible to severe illness and death. Infectious diseases and pandemics easily piggyback on existing health issues to become increasingly lethal.

The coronavirus pandemic has shown that the current population density and behavior of humans render many countries impotent when attempting to isolate or arrest virulent infectious diseases. It also illustrates how easily our current health care infrastructure can be overwhelmed.

Because of COVID-19, it's much easier to envision what could happen with a pandemic accompanied by several natural or human-caused disasters. There is no shortage of fictional books and movies depicting this scenario that are becoming eerily close to a new reality. The inconsistent response to the coronavirus pandemic between countries, regions, and individuals is reflective of human behavior reacting to a threat some consider real and imminent and others consider abstruse.

To satisfy the increased demand for personal protective equipment, testing for the virus, and producing and administering a universally available vaccine, efforts are singularly focused on quickly scaling up manufacturing capabilities. This requires more building, mining, manufacturing, energy development, and deforestation and results in more trash and toxins in the air and water.

The pandemic has resulted in a worldwide surge in medical waste, in particular plastic waste. Wuhan, China, was the epicenter of the coronavirus. The city's medical waste increased from between 40 and 50 tons per day before the outbreak to about 247 tons on March 1, 2020. Other large cities have experienced similar increases. Those countries and cities with inadequate waste disposal infrastructure prior to the pandemic have experienced a significant increase in waste and toxins in their environment.

Campaigns to reduce single-use plastics were temporarily derailed by COVID-19 as a result of hygiene and safety concerns. Some stores wouldn't allow the use of reusable bags amid concerns of them spreading disease. It's now difficult to go anywhere in the United States without seeing discarded disposable gloves and masks in parking lots and along highways. Significant increases in trash, waste, and toxins in the environment in nearly all countries were a byproduct of the human reaction to the pandemic.

The United States government reacted to the economic and human death toll of the coronavirus by suspending the enforcement of all environmental laws for over a month. While never a priority, systemic environmental problems like climate change and biodiversity loss disappeared from the public consciousness during the pandemic as politicians and major media outlets focus almost exclusively on the immediate effects of the disease and its economic implications. Environmental degradation is rarely mentioned even though it may have been the underlying cause for the coronavirus becoming a human contagion.

There are many who now purport to the novel coronavirus being cooked up in a Chinese lab and then escaping through some kind of accident. Whether this is the case may never be known, but it is not unheard of. The virus responsible for SARS sickened a graduate student who worked in a Singapore lab. The presence of laboratories that study potential viruses are a risk in and of themselves. These labs wouldn't exist if it weren't for the risk of virus transmission from animals to humans made possible by environmental degradation. This is just one of many examples that illustrate the real possibility that we will create an environmental apocalypse in desperate last-minute attempts to avert one.

The COVID-19 response is typical of all efforts to deal with human-induced catastrophes. They lead to an ever-increasing snowball effect, creating bigger, more ominous crises in the future. Without exception, our efforts to mitigate catastrophic events like COVID-19, whose underlying cause is environmental degradation, is accomplished by increasing environmental degradation. It is the same verse in a song that is being repeated over and over.

There are no catastrophes where the human reaction isn't accelerating our inevitable demise from the planet. Simultaneously, there is an ever-increasing number and frequency of disasters and emergencies. Everything we do affects everything else. The captains of the human extinction spaceship shift into warp drive when confronted with a crisis.

CATASTROPHES

Natural events have caused species extinctions since the beginning of life on Earth. While the risk at any given time of these events is small, when it happens, the consequences are catastrophic. Asteroid impact and supervolcanoes are among the natural astronomical and geophysical risks to human extinction.

NASA currently believes a supervolcano eruption poses a greater risk to humanity than asteroids. Researchers have found that if a supervolcano did erupt, a "volcanic winter" would ensue that could outlast the amount of stored food worldwide, affecting all continents on the planet.

Comets, supernovas, and gamma ray bursts have been observed in our galaxy. Although their effects would be catastrophically bad, supernovas and gamma ray bursts are extremely rare and unlikely to occur during the Holocene epoch. Comets are also rare but pose a greater risk because of their unpredictability.

Natural extinction risks for all species are now dwarfed by the extinction risks from human activities. The likelihood of any organism disappearing forever are amplified where multiple risk factors and events conspire to increase vulnerability. The combination of natural and human-caused risks conspires to increase extinction risks for all complex organisms.

Several hundred years ago, intact forests and wetlands could have mitigated many of the changes to the global climate resulting from potential natural events like supervolcanos and smaller meteor impacts. The natural ability of the planet to mitigate these more significant natural events is now so compromised that the likelihood of species surviving is becoming increasingly remote.

DESERTIFICATION

Mentioning the word *desertification* usually brings visions of the landscapes depicted in the *Mad Max* movies, deserts, and wastelands. The reality is desertification is the human-caused land degradation in areas with low or variable rainfall known as drylands. These lands account for more than 40 percent of the world's terrestrial surface area. In these marginal areas, human activity can easily stress the ecosystem beyond its tolerance limit, resulting in degradation of the land. The presence of a nearby desert has no direct relationship to desertification. As a result, most people don't clearly understand the concept.

Scientists are beginning to define *desertification* more consistently as a catch-all term for land degradation in water-scarce parts of the world. As a result of degradation, the land can no longer support the same plant growth it had in the past, the change being permanent on a human timescale. Climate change exacerbates desertification by increasing the risk of drought. Conversely, desertification can exacerbate climate change by reducing carbon sequestration and reducing evapotranspiration in plants.

Photosynthesis supports not only humanity but all animals. The thin, fragile layer of arable soil covering the planet supports those plants providing complex organisms with oxygen and food. The abuse we heap on this delicate layer of the Earth's surface knows no limit. Degradation is driven by numerous factors including plowing, grazing, mining, highways and other roads, powerlines, wind turbines, buried pipelines, solar farms, urbanization, and deforestation. Trees and other vegetation are cleared away, animal hooves compact the soils, and crops deplete nutrients in the soil. Chemicals used in agriculture and toxic waste are dumped onto it. The result of these activities is less productive soils no longer capable of retaining water and regrowing native plants.

According to the European Commission's World Atlas of Desertification, 75 percent of the planet's land area is already degraded. The consequences of desertification include the loss of biodiversity; an inability of the land to support animals, plants, or people with resulting malnutrition; respiratory disease caused by dusty air; and other diseases stemming from a lack of clean water.

Desertification has laid claim to millions of acres in North America and is representative of what has happened across the planet. The majority of North America's dryland areas are found in the high western plains and the lower slopes of the mountains that rise above them, a land born of the great uplift of the Rocky Mountain system many millions of years ago. The climate is often unpleasant, vacillating between long winters when blizzards drive down from the mountains and snow lies deep on the plains to summers where heat is relieved by only minimal rainfall. Drought and drying winds steal moisture from any plant attempting to survive.

Sagebrush was the result of long ages of evolutionary experimentation in developing a plant capable of surviving this hostile environment. Low-growing and shrubby, it could hold its place on the mountain slopes and plains and, within its small gray leaves, defy the desiccating winds. It was the most important plant over a huge expanse of western North America, providing the foundation for hundreds of species including sage grouse (described by Lewis and Clark as the "cock of the plains"), pronghorn antelope, and mule deer.

In a land that nature found suited to bunchgrasses mixed with and under the sage, domestic livestock first wiped out most of the native grasses over a period of approximately 75 to 90 years after the Civil War. Cattle don't find sage palatable, and cattlemen didn't have a remedy for wiping out the tenacious sagebrush until the advent of chemical spraying.

To satisfy the insatiable demands of the stockmen, beginning in the 1950s for a quarter of a century federal agencies plowed, sprayed, and killed millions of acres of sagebrush, planting grass monocultures for livestock. They ripped apart the fragile life on this continent's drylands, contributing in their uniquely American way to desertification.

In these naturally dry areas, climate change is affecting the frequency and magnitude of extreme events like drought and floods. Already under extreme stress from the high numbers of livestock, the land has no capacity for resilience. As plants die off due to lack of water, the soil becomes bare and is more easily eroded by wind and by water when the rains do eventually come.

The last 150 years of overgrazing on western North American rangelands and chemical control of sagebrush have resulted in the loss or declining quality of vegetation; the compaction, hardening, or loss of the soil; an increase in wildfires as a result of annual invasive grasses replacing native bunchgrasses; and declining water tables from overgrazing meadows and areas along streams and rivers.

Fifty years ago, the ubiquity of overgrazing by livestock was easily the most significant contributor to desertification. While its effect has worsened, myriad other activities are now attempting to leapfrog overgrazing in significance. Oil and gas development, fracking, wind farms, pipelines, powerlines, roads and highways, and urbanization are a few of the many factors contributing to desertification on Earth.

FOOD AND WATER

There are no human-created effects on the planet that don't also affect available clean water. Many areas of the world no longer have the fresh water to meet the needs of rapidly increasing populations and urbanization. One out of 10 people on the planet currently lack basic drinking water access. Every day, more than 800 children under the age of five die from diarrhea attributed to poor water and sanitation. Fresh, clean water is vital to our existence. There is little advantage to having access to abundant water that makes a person sick or kills them over no water at all. Populations are willing to go to war to ensure adequate clean water supplies.

There are already conflicts occurring between countries in the Middle East and Africa that can be attributed to available water. Water stress can also trigger destabilizing secondary effects, which has and will continue to lead to ever-increasing conflicts in the form of terrorism, insurgencies, civil unrest, and localized violence. Water stress can empower violent extremist organizations and place stable governments at risk. The potential exists for water stress to amplify existing tensions between states and countries.

Until recently, most people in the United States thought of third world countries as the only places lacking access to safe drinking water. The contaminated drinking water in Flint, Michigan, and more recently Jackson, Mississippi; Kiev, Ukraine; Rome, Italy; and Athens, Greece changed the narrative.

Beginning in the 1830s with the first lumber mills, the Flint River served as an unofficial waste disposal site for treated and untreated effluent from the many local industries that had sprung up along its shores. Among these were meat-packing plants, lumber and paper mills, and carriage and car factories. The river also received raw sewage from the city's waste treatment plant, agricultural and urban runoff, and toxins from leaching landfills. The Flint River is rumored to have caught fire on two occasions. Several times recently before lead was positively identified as a problem, people in the city were asked to boil water because of fecal coliform contamination.

For most of its history, Flint, like many other cities, didn't worry about the pollution of its river. As the forests around Flint were gradually cleared, the city switched from lumber and paper mills to carriage factories and finally to automobile factories. In 1893, the city began drawing its drinking and industrial waters from the Flint River and discharging its untreated waste downstream. In the 1930s, fish began to disappear. The once-plentiful walleye population in the Flint River completely disappeared in the 1940s due to polluted water poisoning incubating eggs. As is our custom, the early warnings resulting from the disappearance of fish and wildlife were ignored, and pollution in the river continued.

Now after 200 years of waste and toxins being dumped in the river, Flint is similar to Chernobyl. With lead poisoning and carcinogens in the water, the total health consequences may not be known for years. At one point, there were even calls to abandon Flint and resettle its citizens. For the time being, Flint will continue to muddle along, handicapped by several centuries' worth of environmental abuse. Flint is representative of what will occur with increasing frequency across the planet.

Lack of access to affordable food has proven to trigger revolutions and spark unrest across the world. Regions in the Middle East and sub-Saharan Africa suffer from political unrest and costly humanitarian crises because of an inability to meet their populations' basic food demands.

From 2006 to 2011, a severe multiyear drought affected Syria, contributing to massive agriculture failures. Food was then used as a weapon of war in the Syrian conflict when Syrian President Assad purposely cut people off from humanitarian assistance. The result was 6.8 million Syrian refugees fleeing to other countries.

Economic and political instability will continue to increase as food and water supplies are impacted in more areas on the planet by pollution, drought, flooding, and other extreme weather events. However, while the focus is almost entirely on the supply side of the economic equation, the demand side may be the bigger issue. The planet is currently adding an additional 80 million mouths to feed each year, 4.5 times more people than the entire population of Syria. As the chasm between uncertain supplies and increasing demand widens, the social and biological implications for humanity are both unpredictable and alarming.

PETS

I repeat myself: On a finite planet where a great number of organisms share the same space and depend on the same finite products of photosynthesis, the continuous expansion of one species results in the contraction of others. There is nothing more sacred to humans than our pets. While human overpopulation is occasionally mentioned in passing, pets always receive a free pass.

Dogs and cats in particular are anthropomorphized by their owners who consider them part of the family. Some people become more emotional about negative comments about pets than their own children. Anyone saying something negative about pets is viewed by many as a monster. However much I like and enjoy dogs, cats, and other pets, it doesn't blind me to their effects on the planet.

A few years ago, I was a member of a small gym that had almost none of the conveniences of most modern gyms. The redeeming quality of this gym was the proximity to my home. It had only three single bathrooms with showers for the entire gym membership. The two easily accessible showers in the front of the gym were spartan but came equipped with chairs, shower curtains, and a few hanger hooks for towels and clothes.

The one in the back behind the janitor's storage room only had a chair. It was similar to a prison cell but with a shower instead of a bed. Because of its stark accommodations, it was almost always available when the other two shower rooms were occupied. It not being a desirable shower room resulted in me gravitating there regularly after exercising.

One morning I walked into the room, turned on the light, and caught a glimpse of movement along the wall. Huddled in the corner of my shower room was a terrified mouse. My first thought was *How in the world did you survive the gauntlet of cats?* Any mouse living in an urban or suburban area that has somehow survived the cat densities prevalent nearly everywhere immediately had my respect and admiration. After a 10-minute game of hide-and-go-seek around the toilet, the mouse rodeo ended when I finally corralled it in my T-shirt, wished it well, and sent it back up the pipe to wherever it came from.

In every neighborhood I have lived, the domestic cat densities were astronomical. Cats hang out like gangs of teenagers, killing small mammals and birds out of sheer boredom. The average outdoor pet cat kills around two animals per week.

There are currently 86 million free-ranging pet cats and another 30 to 80 million feral cats in the United States alone. While humans make the initial incursions destroying native ecosystems through urbanization, domestic cats broaden the effect by assassinating all small native wildlife a quarter to half mile from any homes or buildings. Feral cats extend that radius even farther. Cats kill between 1.4 and 3.7 billion birds and between 6.9 and 20.7 billion mammals each year. Cats have already contributed to or are entirely responsible for 33 species extinctions around the world.

Native birds and small mammals are already being negatively affected on multiple fronts by climate change, urban development, deforestation, toxic chemicals, invasive species, and the degradation of native grasslands. Domestic and feral cats are just one more factor making their survival increasingly tenuous.

The effects on our environment from the loss of the multitude of small mammals and birds extend well beyond their direct killing by domestic cats. Birds are important for the pollination and seed dispersal of many plants. Small mammals are prey for bobcats, foxes, dingoes, weasels, caracals, ocelots, snakes, and birds of prey. Domestic cats have essentially replaced all those animals in urban and rural areas.

American dogs and cats consume the same number of calories as the entire human population of France in a year. The meat consumption by dogs and cats creates the equivalent of about 64 million tons of carbon dioxide a year, the equivalent of a year's worth of driving from almost 14 million vehicles.

America's pets produce about 5 million tons of feces in a year, as much as 90 million Americans. Where humans have developed wastewater treatment facilities to prevent our feces from contaminating the water we use for drinking and recreation, pet feces are often ignored and wash into streams, rivers, ponds, marshes, and lakes.

Many public access areas for recreating near rivers, streams, and lakes have restrooms to try to prevent people from crapping along or near waterways. On a typical day, more pet owners will drive into these areas and let their dogs out to run around and relieve themselves than people actually using the restroom. When it rains, fecal coliforms from the dog poop are eventually washed into lakes and rivers. Downstream, irrigators either pump or divert water out of the river to irrigate crops. It should come as no surprise that people occasionally get sick and even die from ingesting lettuce and other crops contaminated with fecal coliforms.

Many of these same people would religiously pick up their dog's feces if it were in their yard or their neighbor's yard. The lack of appreciation about their pet's contribution to deteriorating water quality and even human health is partly a reflection of how much people take clean water for granted. The perception a pet can do no wrong also contributes significantly to the problem. A pet owner's own crap requires a restroom to prevent water contamination, but even their pet's shit is pure.

Pet feces can carry diseases that make water unsafe for contact. These include the bacterial infections campylobacteriosis and salmonellosis; toxocariasis, a roundworm infection; and toxoplasmosis and giardiasis, both protozoan parasite infections. Should our water quality infrastructure crash as a result of either natural or human-caused disasters or civil unrest, our pets' feces in public waterways may have a direct effect on our immediate survival. Dogs and cats will continue to contribute in a meaningful way to global warming, declining water quality, and the loss of biodiversity on the planet.

Reverse engineering the items we need for our pets further illuminates the negative effects on natural ecosystems. There are three mines in Wyoming necessary just to supply the majority of the material necessary for kitty litter. There are approximately 4 billion pounds of cat litter sold every year in the United States.

While the care many pets experience from their owners rivals what their own children receive, there is a much-darker-minded pet owner who mistreats, abuses, and when inconvenienced, abandons pets. Living on a small ranch, my family was often the recipients of dogs and cats left abandoned on or near our property. My mom, sisters, and I made many sad trips to animal shelters with dogs and cats that were almost certainly going to be euthanized.

This same class of cruel and thoughtless pet owner is responsible for the green iguana and Burmese python invasions into natural habitats. Like the cats and dogs released near our ranch, when the pet is no longer wanted or becomes too difficult or expensive to care for, they are released into rural areas. Many pets die in this manner, but those that don't create major environmental problems.

The python population in southern Florida is somewhere between 30,000 and 300,000. Pythons are having devastating impacts upon native animals. Declines in raccoon, opossum, bobcat, rabbit, and fox have been documented and are likely a direct result of the increased population of pythons.

It takes approximately two acres of land to keep a medium-sized dog fed. In 2004, the average citizen of Vietnam had an ecological footprint slightly less than that. For an Ethiopian, only two-thirds the amount of land is needed.

Many rationalize the impact of pets, particularly dogs and cats, as a part of nature. Nature, in the broadest sense, is the natural, physical, and material world. Technically, chainsaws, bulldozers, and pets are a part of nature, but none are part of native ecosystems. Each is uniquely destructive. However much we like to bury our heads in the sand as to the effects from pets, we live in a world with finite resources.

INVASIVE SPECIES

While the human population is growing at an exponential rate, we aren't unique. There are many species that have recently experienced similar exponential growth, almost entirely because of human intervention. Knowingly in some cases and unwittingly in others, people have moved species from where they were native to where they aren't. Because there are few environmental constraints in their new homes, these species proliferate until they reach a new equilibrium at densities many times higher than where they were native. They often modify or destroy the native ecosystems where they have been introduced.

The twin fungal disasters of the Chestnut blight and Dutch elm disease were described earlier. When it comes to invasive species, fungi are just the tip of the iceberg. In 1935, 101 cane toads were shipped from Honolulu to Australia with the intention of utilizing them in the battle against beetle grubs, which were affecting the region's cash crop—sugar cane. The offspring of these cane toads were intentionally released into the region's rivers and ponds. Cane toads, native to South and mainland Central America, can grow to prodigious sizes. Adults average four to six inches in length, with the largest over nine inches.

Since their release, cane toads have expanded their range, gradually colonizing new areas. By 2005, they were present on nearly 25 percent of the continent. Because cane toads are poisonous, consumption by native animals is usually fatal. The populations of many native species have crashed due either to poisoning from cane toad consumption or in the case of smaller organisms, predation by the toad. Northern quolls, freshwater crocodiles, monitor lizards, snakes, native frogs, and some species of birds have all been negatively affected. Attempts are currently underway to genetically modify the toad so it is less poisonous. Only time will tell whether this solution itself becomes a new threat in an unforeseen way. The law of unintended consequences is a frequent visitor with human efforts at managing ecosystems and species.

The black rat may have been one of the first invasive species to ever be inadvertently moved by humans. The species originated in tropical Asia but is believed to have reached Europe by the first century CE before spreading across the world, hitching rides en masse on European ships. Since then, the black rat has thrived in just about every region of the world and has adapted well to rural, urban, and suburban environments alike.

They do extraordinarily well living largely off the resources associated with human activity. Unfortunately, its success as a species, in combination with the success of numerous other species of invasive rats, is believed to have come at the expense of dramatic population declines and the extinction of countless birds, reptiles, and other small vertebrate species the world over. New Zealand has been particularly hard hit from the effects of black rat infestation. The country is currently engaged in large-scale rat control programs to conserve native bird species.

Black rats thrive in tropical regions but have been largely driven out of more temperate regions by Norway rats which are more successful in colder climates. While rats are widespread, they largely avoid mature forests. Large trees provide little low-level cover and those forest floors are covered with leaves, which are noisy for rats to run across, attracting predators. Logging in rainforests has helped black rats infiltrate those areas because of having dense low-plant regrowth, providing better security for scuttling along on the ground.

For most species, the new environmental constraints imposed after being moved will still be sufficient to prevent them from utilizing all the resources on which they depend. However, available evidence suggests that all species would use whatever resources are at their disposal if they could figure out a way.

Humans are following a similar pattern as these two species. We will continue to grow, reproduce, and occupy new areas until environmental constraints prevent further expansion. The difference between humans and these two invasive species is the scale of our habitat manipulation and environmental constraints. When the human population reaches the threshold where growth is no longer possible, the ecosystems that were once suitable for diverse life-forms will have degenerated to a point they can no longer sustain humans and most other complex, multicelled organisms.

There are thousands of invasive plant and animal species making significant contributions to the loss of biodiversity on the planet. Some invasive species are globally disruptive. Zebra mussels are causing ecological imbalances in rivers and lakes in North America and Europe. Water hyacinth outcompetes native vegetation in the Pantanal wetlands in South America, Kruger National Park in Africa, and the Sundarbans mangrove forests in Asia. European carp have negatively affected aquatic ecosystems in North America and Australia. Invasives will increasingly be the death knell for species and ecosystems struggling to survive the threats described earlier in this chapter.

CHAPTER 10

THE FINAL CHAPTER

Evolution, ecosystem feedback loops within and between threats, and anthropocentrism make human extinction both imminent and inevitable. The basic tenants of evolution are to survive, reproduce, and grow. Humanity, like all species, is helpless to overcome these innate behavioral drivers. No species has the capacity to voluntarily stop reproducing, being constrained either physiologically or by environmental factors. The human population will continue to grow until either using up available resources or poisoning ourselves with the waste products we generate.

Feedback loops can be either positive or negative, depending on whether they speed up or slow down a trend. Positive climate change feedback loops, for example, include the heat island effect from pavement and buildings absorbing solar radiation, more ice melting from glaciers and polar regions reflecting less solar radiation back into the atmosphere, and increasing ocean temperatures that absorb less carbon dioxide. There are now positive feedback loops within feedback loops, continuously amplifying the effects of all threats mentioned in chapter nine.

Anthropocentrism fuels a disinterested, materialistic human populace. Most people believe any problem created is fixable with new or existing technologies. Planetary ecosystems are quickly unraveling, losing their resiliency and ability to buffer human and naturally caused catastrophes. Because most people do not view themselves as part of Earth's ecosystem, complex organisms disappearing from the planet aren't considered important to our long-term persistence. These attitudes are dangerously naive and untrue.

Anthropocentrism also leads some to believe that our supreme importance on the planet means humans bear sole responsibility for the deterioration of our biosphere. This perspective ignores that humans are the product of an evolutionary process and have been influenced by species like leopards, lions, eagles, grasses, and countless other plants and animals for millions of years. It fails to acknowledge the role of chaos and coincidence inherent in all ecosystems, often stemming from climatic and geological changes. Even though humans are precipitating the sixth mass extinction, we are merely puppets of evolution and marionettes of chaos. *Homo sapiens* is just another species.

Time and evolution follow separate paths that never intersect. Evolution is a continuous process that enables species to thrive for varying durations, ranging from several centuries to millennia or even millions of years. Ours will be a short tenure.

There is no silver bullet or magic spell that will save humanity. Carbon taxes, biodiversity initiatives, recycling programs, green energy, fuel-efficient vehicles, and water conservation measures are well meaning, but at best would only delay the inevitable in some minuscule way, if at all.

Most people live in cities and know scarcely anything about the native plants, wildlife, or the planet that supports every aspect of their life. If we don't know anything about nature and biodiversity, then we don't have the capacity to value it. Our universe is limited to what we see. For some, it is a grocery store, department store, restaurants, and concrete jungles. For others, it is dogs, cats, chickens, pigs, ducks, and cows. Our sensitivity extends only toward them. It's like looking through a keyhole. The world is bigger than that and far more complex, whether someone accepts it or not.

Our planet won't make any distinctions between national boundaries, political parties, religious affiliations, ethnic groups, or economic and social beliefs. Humanity can't survive the combined effects of climate change, biodiversity loss, deforestation, desertification, mountains of trash, toxins in the environment, and our imagined self-importance.

The ability of the planet to support the existing population of humans is declining rapidly. No one is willing to give up their SUVs and shopping malls, so we ignore the ecological apocalypse that is being documented every day. People will continue to push for economic growth until the planet's capability to support humanity collapses.

The habitat supporting humanity we call Earth has already deteriorated to a point that an implosion resulting in our annihilation could occur at any time. It is impossible to predict when the planet shrugs us off and how our departure scenario plays out. It might be a year from now or 50 years from now. The probability of our demise increases with every passing year. There is almost no chance humanity will know whether the tree seedlings currently being planted under the auspices of sustainability become large, mature trees.

Sports betting is prevalent in most countries on the planet. An over-under bet is a wager where a sports book will predict a number for a statistic in a given game. The bettors wager that the actual number in the game will be either higher or lower than that number. The goal of a sports book is to have an equal value of bets on both sides of the over-under.

Humanity will soon vanish from Earth. It's no longer a matter of if, but when. My current over-under on human extinction is 2055—32 years. A theoretical bet on the over means you believe some humans will still be present on the planet after 2055. Conversely, a bet on the under means we will be entirely absent prior to 2055. As with sports betting, there is an equal chance of each scenario occurring and the over-under can change as humanity nears our inevitable conclusion.

While most choose to ignore all threats, some look at small pieces of larger threats. An even smaller percentage of people view these threats as additive. While more dangerous when considered as additive rather than isolated or partial threats, the reality is they have a multiplier effect on each other. The consequence of the multiplier effect is an extinction time frame that is frighteningly imminent and grossly underestimated. There are too many threats working in unison, in an infinite number of ways, and increasing in magnitude and scope for humans to escape.

My emotional reaction to our imminent extinction has mirrored the five stages of grief and loss: denial, anger, bargaining, depression, and acceptance. Like grief itself, the stages aren't stops on a linear timeline. While often stuck in the denial stage, I would occasionally jump ahead into the anger phase. This scenario played itself out for most of my career. Most people trying to come to grips with what we are doing to the planet are on the same pogo stick I was—jumping between denial and anger.

Moving past the denial phase finally occurred when our incursions on the planet became too obvious and overwhelming to ignore. The majority of humanity will always remain there. The huge investment made in ignorance and myths along with the lack of connection with our habitat has too much inertia to overcome. Riding out our existence to its calamitous conclusion in complete denial is how most people will continue to feel good about themselves for as long as possible.

For those making incursions into the anger phase, the common thought is *Why aren't people doing something about this travesty?* Empathizing with them is easy because of the amount of time I've spent wallowing in anger.

Denial and the third phase of grief, bargaining, are nearly inseparable when it comes to our destruction of the planet. Bargaining takes place within the mind by trying to explain the things that could have been done differently or better and often starts with "If only": *If only I could have convinced the majority of people on the planet that our very survival was contingent on biodiversity and ancient forests.* Part of finally moving past the denial phase is recognizing the impossibility of changing the course of an unknowing, uncaring, or willfully blind populace.

Grief and depression, the fourth phase of grief, was something I did for months before finally transitioning into the final phase of acceptance. I still care too much about wildlife and wild places not to occasionally have bouts of anger and depression, but the time I spend there isn't as prolonged. Similar to how we each deal uniquely with life and grief, every person who recognizes the planet will expel humanity in the near future will have their own journey.

An anecdotal but powerful example of how swiftly the planet is coming unraveled compares the summaries for policy makers provided by the Intergovernmental Panel on Climate Change (IPCC) of 1999, 2014, and 2021. The 1999 summary stated there is a greater need for "[the] understanding of the economic and environmental effects of meeting potential stabilization scenarios (for atmospheric concentrations of greenhouse gases), including measures to reduce emissions from aviation and also including such issues as the relative environmental impacts of different transportation modes."

Fifteen years later in the summary for policy makers, the 2014 IPCC report stated, "Continued emissions of greenhouse gases will cause further warming and long-lasting changes in all components of the climate system…Limiting climate change would require substantial and sustained reductions in greenhouse gas emissions…"

In 2021, the IPCC report stated, "Global surface temperature will continue to increase until at least the mid-century under all emissions scenarios considered. Global warming of 1.5 degrees Celsius and 2 degrees Celsius will be exceeded during the twenty-first century unless deep reductions in CO_2 and other greenhouse gas emissions occur in the coming decades."

While they were still writing like the bureaucrats and scientists they are, comparison of the changing mood of these three statements over a 20-year period shows the members of the IPCC are becoming increasingly desperate and alarmed. Their tone has changed from an almost casual curiosity in 1999 to one of distress and urgency today. The trend of increasing despair will continue as they realize governments and individuals are incapable of making the changes necessary to save humanity. There is now no amount of change humanity can make that isn't too little, too late.

Between 1980 and 2022, Earth's human population increased by over 80 million people annually. The planet continues to add millions more vehicles on roads and improve transportation infrastructure with more and wider highways. Deforestation, logging, and wildfires on the planet soared to levels never before witnessed. The amount of trash and toxins generated increases every year. Overgrazing and desertification by domestic livestock and urbanization are continuing unabated. The deadly coronavirus pandemic has disrupted nearly everyone's life on the planet and appears to be a virus that will continually mutate and become a long-term problem. The number of extinctions and species in danger of disappearing increases every year. With all that has happened, it represents but a small taste of things to come.

Great numbers of people are already suffering from the effects of biodiversity loss and climate change, but the root cause is hidden under the backdrop of social, health, and economic disasters. The extraction of natural resources can't stop or even slow down. The only thing we know how to do is to grow. We will continue to grow the economy, grow our families, grow our businesses, and strive for success every year until human existence on the planet ceases.

There were times I believed we were an evolutionary mistake, but that implies evolution is something other than a process. We aren't a mistake. *Homo sapiens* is an evolutionary dead end. The question isn't whether we will soon be absent from the planet, but what Earth will look like when the evolutionary process kickstarts itself after our departure.

Microbial life has been present on Earth for more than 3 billion years. Microbes run the world. Microbes occupy the full range of environments from bedrock and deep ocean bottoms to the upper atmosphere, covering environments so inhospitable to life that we can hardly make them worse. At minimum, microbes will still be here when we take our permanent leave of absence. It is my fervent hope that there are a number of plants, invertebrates, fungi, and vertebrates that are even more resilient than humans that may also survive. Cockroaches, rats, and jellyfish have positioned themselves to seize the reins. There is some consolation in knowing we will not eradicate all life on Earth.

With only six known major extinction events on the planet, the sample size is too small to know whether one out of every six extinction events is normally precipitated by evolution itself. It's a question with an answer we'll never know.

Arrogance, along with our unfailing belief in human superiority and intelligence, creates the illusion we'll discover a new technology allowing us to persist or enact countermeasures to ameliorate our ongoing attacks on the planet in time to prevent our demise. Both are pipe dreams. Our primary strategy for dealing with an impending catastrophic environmental crisis is ignoring the evidence, crossing our fingers hoping it won't happen, or laughing it away.

Equally unrealistic are those who have convinced themselves that all that is necessary are small changes in peoples' lifestyles. Entire books have been written about the topic. While this concept may have validity for health and social issues, it has no realistic basis when examining the ecological status of the planet.

There is an infinite number of variables that will come into play as humanity disappears from the planet, including ecological, societal, economic, and natural disasters. Even though deteriorating ecosystems will drive our extinction and remain unrecognized to most people, my best guess is all of these will interact with each other in some fashion as our final act plays itself out.

While our lives are an insignificant speck in the galaxy, they are still a gift to treasure. Armed with this information, everyone will make their own decisions about the best way to move forward with their lives. There is no right or wrong way. Most will opt for continuing with their current life path. Because of having a little more experience in being armed with this information, I will briefly share a few insights that have helped me that you may or may not find useful.

I still maintain my pro-environmental lifestyle, not because my way of living will make a difference but because it helps assuage my guilt at my own contribution to the problem. I'm still deeply saddened humanity will be taking most of the plants and animals with us when we punch the clock the final time.

Unfortunately, most people believe the only way we can do something for the environment is to make a huge lifestyle change, purchase an expensive hybrid vehicle, or donate to an environmental group. Since everything we eat or do has some type of negative effect on the planet, undertaking something positive for the environment every day is ridiculously easy. Use less: less paper products, less plastic, less oil and gas, less water, less anything. Many people in modern society could eat less, not only improving the environment but their own health.

Creativity and imagination can be utilized to minimize natural resources utilized and the amount of waste, emissions, and toxins generated. A few examples include using half a paper towel instead of an entire paper towel, taking a three-minute shower rather than a four-minute shower, or only using the amount of napkins, toilet paper, or tissues you need rather than what is convenient and easy. There is an infinite number of actions people can take to slow deforestation, climate change, desertification, loss of biodiversity, and every other insult we are inflicting on the planet. Truly appreciating the sacrifice other organisms make for everything we do should be motivation enough.

Using less ultimately results in less trash and waste, the goal for anyone concerned about the environment. Trash and waste are the end product of a person's lifestyle choices. The fewer items utilized and the more belongings recycled, repaired, and reused, the less the impact on the planet.

All these personal actions to protect the environment will, at best, add a brief time to our tenure on the planet, even if millions of people adopted them. Any or all positive effects inspired by those who do act will be immediately gobbled up the billions of people who couldn't care less. These conservation actions are solely for those who have accepted our extinction but would prefer to have less responsibility. In the end, they will make little or no difference.

Practicing mindfulness is important for my sanity. While it sounds woo-woo, at a basic level, mindfulness is about remaining in the present moment. This effectively prevents the anxiety of worrying about what will happen in the future and the depression of thinking about what took place in the past. Like our lives, now is also a gift. Fly-fishing oceans, lakes, streams, and rivers or practicing complex movements and exercises force my mind into the present moment. Savoring moments of beauty and quiet help me find some peace: the chipmunk chattering in the pines, the stars and moon in the eternal sky, the grasses waving in the wind, the rising and falling pulse of ocean waves.

Reading novels or watching movies and sporting events provides mental vacations when I'm unable to be involved in more active pursuits. What works for me may not work for you; experiment until you find something that does. Mindfulness will likely look a little different for nearly everyone and can help anyone no matter what their current mental challenges.

Humor is important. Reading comic strips every day, trading humor with friends on the internet, and watching clips of the late-night television shows are ways I incorporate humor into my life. Be willing to laugh at yourself and the world around you.

I attempt to balance mindfulness with being aware. Staying apprised of international and national developments is important but does make me anxious. The reality of our situation is driven home daily as much by what is not said on media outlets as by what is. In spite of the environment driving economic growth, our longevity on the planet, and everything we do, it is not a priority and never will be.

Embrace the majestic stream of life, which includes pain and suffering as well as happiness and joy. Resist the temptation to declare that all evil exists within *them* and all goodness inside *us*. This tome is for those with the capacity to see the truth of the planet's woes and find consolation in knowing our departure won't prevent the future existence of other new and interesting life-forms thousands and millions of years from now. This book has everything to do with removing the cloak of escape and of self-deception and embracing a hard-won, steady, and whole perception of reality.

My contingency plan for dealing with what could be chaotic, dangerous situations when our society and ecosystems become completely unraveled is the same that reputable self-defense instructors give. The best chance of surviving conflict is avoiding it for as long as possible. Take a lesson from our ancient hunter-gatherer societies. If hard-pressed by stronger rivals, it may have been difficult and dangerous, but they were usually able to move on. Farmers, forced to stay put to protect houses and fields, fought to what was often the bitter end. You could be the most badass person in the entire world, but if you are involved in a conflict, your chance of survival is much less than anyone who avoids it.

We're all mortal, but for many, it is much more imminent and real than for others. The lesson we learn from people dealing with terminal illnesses is to enjoy your life now. There is no guarantee of a tomorrow. While that has always been the case for individuals, it is now the case for our species. While we may seem so terribly important, humanity's presence on the planet will barely register as a hiccup on the cosmic timescale. Adopt humility.

There is an increasing amount of evidence that the use of psychedelic drugs like psilocybin and LSD have helped terminally ill patients deal with the anxiety and depression associated with their mortality. For those with a heightened awareness of the planet's limitations, these drugs' potential for providing relief from the mental aspects of an impending environmental apocalypse may be similar to that for the terminally ill.

Published in 1962, Rachel Carson's *Silent Spring* was the springboard for the modern environmental movement. It focused on environmental problems caused by synthetic pesticides and led to a nationwide ban on DDT for agricultural uses. The ban led to a rebound in populations of bald eagles and peregrine falcons and was considered a major environmental success. In a letter to a friend, she wrote, "The beauty of the living world I was trying to save has always been uppermost in my mind—that, and anger at the senseless, brutish things that were being done…now I can believe I have at least helped a little."

Similar to Rachel Carson's, my anger at the senseless, brutish things that are being done to all things wild was one of the primary motivators for writing this book. Trying to preserve the integrity of this amazing world has always been foremost in my enthusiastic, dedicated, and usually failed efforts.

If Rachel Carson were alive today, she would be horrified to find the oceans she loved have been turned into a trash dump and the biodiversity of the planet is being flushed down the proverbial toilet. Even the insecticides highlighted in *Silent Spring* have now become more sophisticated, pervasive, and dangerous to biodiversity. The ban on DDT was followed by a tenfold increase in the use of pesticides in agriculture. She would be dismayed to see that the net result of her effort would not even register as a speed bump on the road to global catastrophe.

Writers who address the abuse we've inflicted on our planet tend to close on a hopeful note, offering reassuring platitudes and false hope that a few tweaks and some minor lifestyle modifications and all will be well. I can't in good conscious make those same assurances. Our situation is dire. Humans will make the lemming suicide myth created by Disney in the 1950s our own reality, dragging most of the planet's complex organisms over the cliff with us.

While my efforts to stop humanity from destroying itself were paltry compared to the 1,700 environmental activists murdered around the globe in the last decade trying to protect biodiversity, I gave it my best effort. Fighting against progress, freedoms, and wealth—what most perceive as our race's birthright is not something society tolerates.

People have now made understanding and appreciating the essential services that forests, grasslands, wetlands, and oceans confer to our ability to persist on the planet unnecessary. This book is meant to foster realistic expectations of the capabilities of both us and our complex and finite planet. There is no need to fly to the moon to experience incredible marvels. All that is necessary is to walk outside and truly appreciate the intricacies of Earth and its many and varied inhabitants.

Our separateness from nature is delusional. This delusion creates a virtual prison, making us incapable of widening our circle of compassion to embrace all living creatures and the whole of nature in its beauty. In the end, it will be our undoing.

The first hominids appeared on the planet approximately 6 million years ago. We were just another species for 97 percent of our time on Earth. We've proliferated with stunning speed, populating every habitable environment on Earth while creating an incredibly complex web of technology to support our needs and wants. In so doing, we are destroying the planet's life-sustaining natural processes. The culmination of our 200,000 years of meteoric ascendancy will be *Homo sapiens* vanishing forever.

Some will view this effort as a failure by not providing a strategy to prevent our demise. Walking back several million years of evolution isn't possible. This book provides the rationale for resignation and acceptance. We are a species poised to soon disappear. Humanity is quite literally…racing to extinction.

Live in each season as it passes, breathe the air, drink the drink,
taste the fruit, and resign yourself to the influence of the earth.

—HENRY DAVID THOREAU

ACKNOWLEDGMENTS

I'd like to thank Lisa Wilkinson, Tom Krauel, and Betsy Krauel for providing their unflagging support and encouragement, insightful questions, unique viewpoints, helpful comments, and most importantly reassurance for a neophyte writer. A special thanks is due to Kristin Thiel, Mark Warren, and Julian Cribb for their willingness to provide meaningful suggestions and advice without ever having met this writer.

Denise Boggs deserves my gratitude for generously supplying research material for this book, beginning long before I made the decision to begin writing. Thanks also are due to Elise Roberts, Jessie Glenn, and Kristen Krauel for their review and help with specific topics that are an integral part of the book.

My life is forever entwined with the struggles of the following imperiled organisms. In their own unique way, they helped shape my life: quaking aspen, Townsend's big-eared bat, bull trout, American pika, spikedace, Canada lynx, greater sage grouse, Allen's lappet-eared bat, Douglas fir forests, northern spotted owl, Rio Grande silvery minnow, Jemez Mountains salamander, sagebrush, and the northern aplomado falcon.

PSEUDO-APPENDIX

WITH ADDITIONAL INFORMATION

CHAPTER 1

Brain, C. K. *The Hunters or the Hunted? An Introduction to African Cave Tophonomy.* The University of Chicago Press, 1981.

Corbett, J. *The Man-Eating Leopard of Rudraprayag.* Oxford University Press, 1948.

Doughty, C. E., J. Roman, S. Faurby, A. Wolf, A. Haque, E. S. Bakker, Y. Malhi, J. B. Dunning Jr., and J.-C. Svenning. "Global Nutrient Transport in a World of Giants." *PNAS* 113, no. 4 (2015): 868–73. https://doi.org/10.1073/pnas/1502549112.

Flannery, T. *The Future Eaters: An Ecological History of the Australasian Lands and People.* Grove Press, 1994.

Malhi, Y., C. E. Doughty, M. Galetti, F. A. Smith, J. C. Svenning, and J. W. Terborgh. "Megafauna and Ecosystem Function from the Pleistocene to the Anthropocene." *PNAS* 113, no. 4 (2016): 838–946. https://doi.org/10.1073/pnas.1502540113.

Patterson, J. H. *The Man-Eaters of Tsavo and Other East African Adventures.* Macmillan, 1907.

Pontzer, H. "Overview of Hominin Evolution." *Nature Education Knowledge* 3, no. 10 (2012):8.

Roser, M. "Measurement Matters—The decline of maternal mortality." Our World in Data, 2017. https://ourworldindata.org/measurement-matters-the-decline-of-maternal-mortality#.

Smith, F. A., R. E. E. Smith, S. K. Lyons, and J. L. Payne. "Body Size Downgrading of Mammals Over the Late Quaternary." *Science 360* (2018): 310–13. https://doi.org/10.1126/science.aao5987.

Stupaczuk, M. *Seize the Prey: Raptor Predation on Primates.* Prized Writing, 2010.

CHAPTER 2

Harari, Y. N. *Sapiens: A Brief History of Humankind.* Harper, 2014.

Pohle, A.-K., A. Zalewski, M. Muturi, C. Dullin, L. Farkova, L. Keicher, and D. K. N. Dechmann. "Domestication Effect of Reduced Brain Size Is Reverted When Mink Become Feral." *Royal Society Open Science* 10, no. 7 (2023). https://doi.org/10.1098/rsos.230463.

Stibel, J. "Decreases in Brain Size and Encephalization in Anatomically Modern Humans." *Brain, Behavior and Evolution* 96, no. 2 (2021): 64–77. https://doi.org/10.1159/000519504.

Suzuki, D. and W. Grady. *Tree: A Life Story.* Greystone Books, 2004.

Wohlleben, P. *The Hidden Life of Trees: What They Feel, How They Communicate—Discoveries from a Secret World.* Greystone Books, 2016.

CHAPTER 3

Audubon. (n.d.). Passenger Pigeon. Audubon.. Retrieved 2019, from https://www.audubon.org/birds-of-america/passenger-pigeon#:~:text=Let%20us%20now%2C%20kind%20reader,there%20was%20little%20under%2Dwood.

Isenberg, A. C. (2000). The Destruction of the Bison: An Environmental History, 1750-1920. Cambridge University Press.

Kolbert, E. (2014). The Sixth Extinction: An Unnatural History. Henry Holt and Company.

Lockwood, J. A. (2005). Locust: The Devastating Rise and Mysterious Disappearance of the Insect that Shaped the American Frontier. Basic Books

Marneweck, C., A. R. Butler, L. C. Gigliotti, S.N. Harris, A.J. Jensen, M. Muthersbough, B.A. Newman, E.A. Saldo, K. Shute, K.L. Titus, S.W. Yu, and D.S. Jachowski. Shining the spotlight on small mammalian carnivores: Global status and threats. (2021). *Biological Conservation* 255. 109005. doi.org/10.1016/J.biocon.2021.109005

CHAPTER 4

Ceballos, G., A. Davidson, R. List, J. Pacheco, P. Manzano-Fischer, G. Santos-Barrera, and J. Cruzado. "Rapid Decline of a Grassland System and Its Ecological and Conservation Implications." *PLoS One* 5, no. 1 (2010): e8562. https://doi.org/10.1371/journal.pone.0008562.

Egan, T. *The Big Burn: Teddy Roosevelt and the Fire that Saved America.* Mariner Books, 2009.

Ripple, W. J., C. Wolf, M. K. Phillips, R. L. Beschta, J. A. Vucetich, J. B. Kauffman, B. E. Law et al. "Rewilding the American West." *BioScience* 72, no. 10 (2022): 931–35. https://doi.org/10.1093/biosci/biac069.

CHAPTER 5

Jensen, D., L. Keith, and M. Wilbert. *Bright Green Lies: How the Environmental Movement Lost Its Way and What We Can Do About It.* Monkfish Book Publishing, 2021.

CHAPTER 7

Cain, S. *Quiet: The Power of Introverts in a World That Can't Stop Talking.* Crown Publishers, 2012.

Heffernan, M. *Willful Blindness: Why We Ignore the Obvious at Our Peril.* Walker & Company, 2011.

CHAPTER 8

Shellenberger, M. Apoc*alypse Never: Why Environmental Alarmism Hurts Us All.* Harper, 2020.

CHAPTER 9

Bergstrom, D. M., B. C. Wienecke, J. van den Hoff, L. Hughes, D. B. Lindenmayer, T. D. Ainsworth, C. M. Baker et al. "Combating Ecosystem Collapse from the Tropics to the Antarctic." *Global Change Biology* 27, no. 9 (2021), 1692–1703. https://doi.org/10.1111/gcb.15539.

Carson, R. *Silent Spring.* Houghton Mifflin, 1962.

Cribb, J. *Earth Detox: How and Why We Must Clean Up Our Planet.* Cambridge University Press, 2020.

Ehrlich, P. R. *The Population Bomb.* Ballentine Books, 1968.

Klein, N. *This Changes Everything: Capitalism vs. the Climate.* Simon & Schuster, 2014.

Kolbert, E. *Under a White Sky: The Nature of the Future.* Crown, 2021.

Law, B. E., T. W. Hudiburg, L. T. Berner, J. J. Kent, P. C. Buotte, and M. E. Harmon. "Land Use Strategies to Mitigate Climate Change in Carbon Dense Temperate Forests." *PNAS* 115, no. 14 (2018): 3663–68. https://doi.org/10.1073/pnas.1720064115.

Peng, L., T. D. Searchinger, J. Zionts, and R. Waite. "The Carbon Costs of Global Wood Harvests." *Nature* 620 (2023): 110–15.

Watari, T., B. C. McLellan, D. Glurco, E. Dominish, E. Yamasue, K. Nansai. "Total Material Requirement for the Global Energy Transition to 2050: A Focus on Transport and Electricity." *ScienceDirect* 148 (2019): 91–103. https://doi.org/10.1016/j.resconrec.2019.05.015.

Wallace-Wells, D. *The Uninhabitable Earth: Life After Warming.* Tim Duggan Books, 2019.

CHAPTER 10

Pollan, M. *How to Change Your Mind: What the New Science of Psychedelics Teaches Us About Consciousness, Dying, Addiction, Depression, and Transcendence.* Penguin Press, 2018.

ABOUT THE AUTHOR

Lyle Lewis is a former endangered species biologist with the US Department of Interior. He is the founder of the Western Bat Working Group and Southwestern Carnivore Committee and has received numerous awards for bat, carnivore, and natural area conservation. His career and love of the outdoors has taken him to many of the wildest backcountry areas in North America.